Plant Life

Plant Life

A Brief History

FREDERICK B. ESSIG

OXFORD
UNIVERSITY PRESS

OXFORD
UNIVERSITY PRESS

Oxford University Press is a department of the University of
Oxford. It furthers the University's objective of excellence in research,
scholarship, and education by publishing worldwide.

Oxford   New York
Auckland   Cape Town   Dar es Salaam   Hong Kong   Karachi
Kuala Lumpur   Madrid   Melbourne   Mexico City   Nairobi
New Delhi   Shanghai   Taipei   Toronto

With offices in
Argentina   Austria   Brazil   Chile   Czech Republic   France   Greece
Guatemala   Hungary   Italy   Japan   Poland   Portugal   Singapore
South Korea   Switzerland   Thailand   Turkey   Ukraine   Vietnam

Oxford is a registered trademark of Oxford University Press
in the UK and certain other countries.

Published in the United States of America by
Oxford University Press
198 Madison Avenue, New York, NY 10016

© Oxford University Press 2015

All rights reserved. No part of this publication may be reproduced, stored in
a retrieval system, or transmitted, in any form or by any means, without the prior
permission in writing of Oxford University Press, or as expressly permitted by law,
by license, or under terms agreed with the appropriate reproduction rights organization.
Inquiries concerning reproduction outside the scope of the above should be sent to the
Rights Department, Oxford University Press, at the address above.

You must not circulate this work in any other form
and you must impose this same condition on any acquirer.

Library of Congress Cataloging-in-Publication Data
Essig, Frederick B.
A brief history of plant life / Frederick B. Essig.
    pages cm
Includes bibliographical references and index.
ISBN 978-0-19-936264-6
1. Plants—Evolution. I. Title.
QK980.E85 2015
581.3'8—dc23
2014028846

1 3 5 7 9 8 6 4 2
Printed in the United States of America
on acid-free paper

This book is dedicated to those who have encouraged, guided, and supported my botanical journey: my father, who planted fruit trees and took me backpacking in the mountains; my mother, who let me out of her sight for a month to backpack around Hawaii after high school; my grandmother, who supplied me with books about plants and introduced me to the wild places of Cape Cod; my high school teacher, Bert Hunt, who showed me wild places from the Grand Canyon to the northwestern Cascades; my children, for tolerating frequent wildflower stops on family vacations; and last but not least my wife, Yau-Ping, who tolerated my many absences from home to botanize around the world and who provided encouragement and support throughout this project.

CONTENTS

Acknowledgments ix
Artwork Sources xi
Introduction xiii

1. The Origin of Photosynthesis 1
2. Eukaryotic Plant Life 17
3. Plants Invade the Land 47
4. Vascular Plants and the Rise of Trees 73
5. Seeds and the Gymnosperms 95
6. Darwin's Abominable Mystery 117
7. Adaptations for Pollination and Seed Dispersal 137
8. The Dicotyledonous Grade 167
9. The Monocots 199

Epilogue 231
Notes 235
Glossary 237
Bibliography 249
Index 255

ACKNOWLEDGMENTS

I would like to thank all of those who read and commented on portions of the manuscript during its development: Gordon Fox, K. T. Scott, Rebecca Pinkerton, Monica Metz-Wiseman, and Genevieve Essig. Special thanks go to Merilyn Burke and Andrew Smith for invaluable help in deciphering copyright law and tracking down copyright owners. I thank also the families of William H. Brown and Gilbert M. Smith, particularly Dr. Jennifer Brown and Mrs. Susan Pellet Minamyer, for granting liberal use of drawings from the classic botanical textbooks of these authors, and all the others who granted permission to use illustrations from their own or family member's botanical works.

ARTWORK SOURCES

This book describes the evolution of plant structure, a task that would be virtually impossible without the inclusion of a large number of illustrations. In this day and age, one might turn to color photography for this task, but for the scale of this project, that would have been prohibitively expensive. Botanical line art can achieve the same goals, quite often with superior results. In the hands of a skilled artist, details of plant structure can be portrayed accurately, clearly, and surprisingly lifelike.

The rich tradition of botanical art goes back centuries, developed and refined when photography was not an option, or when photos could be published only in black and white with mediocre quality. One of the goals of this book was to draw upon and showcase this tradition with some of the best line art ever published. Only when nothing was available for a particular topic have I resorted to my own artistic effort, and failing that, to include a few black-and-white photographs. For the majority of topics, however, superb illustrations were available.

I have made every effort to determine the copyright status of each of the illustration sources and to track down the current holder of the rights. If I have made any errors or omissions in that regard, I extend my apologies to the appropriate party and will make the appropriate amends.

The cited sources of artwork and photographs are listed below:

Awramik, S. 1992. The oldest records of photosynthesis. Photosyntheis Research 33:75–89.
Barton, W. P. C. 1818. Vegetable Materia Medica of the United States. H. C. Carey & Son: Philadelphia.
Beck, C. B. 1962. Reconstruction of Archaeopteris and further consideration of its phylogenetic position. American Journal of Botany 49:373–382.
Brown, W. H. 1935. The Plant Kingdom. Ginn & Co.: Boston and New York.
Coulter, J. M., C. R. Barnes, and H. C. Cowles. 1910. A Text-book of Elementary Botany for Colleges and Universities. American Book Co.: New York.

Flora of Pakistan, 2008. http://www.efloras.org/object_page.aspx?object_id=115716&flora_id=5.

Ganong, W. F. 1916. A Textbook of Botany for Colleges. MacMillan: New York.

Grownative, www.grownative.org, a program of the Missouri Prairie Foundation

Gray, A. 1879. Gray's Botanical Textbook, vol. 1. Structural Botany. Iveson, Blakeman & Co.: New York and Chicago.

Hartog, M., 1906. In: S. F. Harmer (ed.). The Cambridge Natural History vol. 1. Protozoa, p. 9. MacMillan Co.: London.

Haupt, A. 1953. Plant Morphology. McGraw-Hill: New York.

Kerner von Marilaun, A., English edition by F. W. Oliver. 1895. The Natural History of Plants. Henry Holt & Co.: New York. (cited as Kerner & Oliver).

LeMaout and Decaisne. 1876. A General System of Botany. Longman's, Green & Co.: London.

Masclef, A. 1891. Atlas des plantes de France. Paul Klincksieck: Paris.

Mauseth, J. 2014. Botany, 5th Ed., Fig. 23.1. Jones & Bartlett. Burlington, MA.

Oltmanns, F. 1905. Morphologie und Biologie der Algen. Gustav Fischer: Jena, Germany.

Raven, P. H., R. F. Evert, and S. E. Eichhorn. 1999. Biology of Plants, 6th Ed. W. H. Freeman & Co.: New York.

Sachs, J. 1874. Lehrbuch der Botanik. 4th ed. Wilhelm Engelmann: Leipzig.

Seward, A. C. 1917. Fossil Plants: For Students of Botany and Geology. University Press: Cambridge.

Smith, G. M. 1938. Cryptogamic Botany. McGraw-Hill: New York.

Sturm, J. G. 1796. Deutschlands Flora in Abbildungen. C. G. Preuss: Nürnberg.

Thomé, O. W. 1877. Textbook of Structural and Physiological Botany. John Wiley & Sons: New York.

Thomé, O. W. 1885. Flora von Deutschland, Osterreich und der Schweiz. Friedrich von Zezschwitz: Gera, Germany.

Transeau, E. N., H. C. Sampson, and L. H. Tiffany. 1940. Textbook of Botany. Harper and Brothers: New York.

Wikimedia Commons (http://commons.wikimedia.org/); posted images licensed by Creative Commons (http://creativecommons.org/licenses/by-sa/3.0/legalcode).

INTRODUCTION

Charles Darwin described the evolutionary origin of flowering plants as an "abominable mystery." According to the Merriam-Webster dictionary, the word "abominable" encompasses just about any unpleasant adjective you can think of: hateful, disgusting, disagreeable, etc. Apparently the scientific world felt this way about the hairy man-apes reported from the Himalayas, but what was it about the flowering plants that evoked such an outcry of annoyance from the father of evolutionary theory?

Darwin built his life and his career around the idea that organisms gradually changed over time and that this was documented by the fossil record. Yet, the flowering plants seem to have appeared rather abruptly during the latter part of the dinosaur era, with the oldest fossils already diverse and not much different from recognizable modern forms. What earlier group of plants had they come from? Where was the fossil record of the intermediate stages of their evolution?

Under the relentless probing of modern paleobotanists and plant geneticists, the mystery is beginning to yield, but the story is far from complete. Several groups of ancient seed plants that could logically have given rise to modern flowers have been discovered, and fossilized flowers more than 100 million years old have been found in China. Together they give us some clues as to what the oldest flowering plants might have been like. DNA analysis has identified the most archaic of living angiosperms, and we can integrate some of their characteristics into our model of early angiosperms as well. The gaps have narrowed somewhat, but the transition between nonflowering seed plants and the earliest known angiosperms is still a large void.

Though the origin of the flowering plants is the most famous of abominable mysteries, it is not the only one. The origins of photosynthesis, and indeed of life itself, are mysteries, as are the origin and early evolution of land plants and the first plants with seeds. Abominable mysteries both large and small abound in the evolutionary history of plants. Each major breakthrough that led plants into a new way of life was preceded by a long period during which whatever was happening left no record. Each breakthrough was also the result of overcoming some environmental limitation or challenge, which opened the floodgates of opportunity for the successful organisms.

We cannot blame holes in the fossil record entirely on the softness of plant tissues, though the hard skeletons of animals have certainly been a help in documenting their history. We have an abundant record of the first plantlike organisms, the cyanobacteria, going back over 3 billion years, for they left clear impressions of their cell walls. Hard evidence for what came before them, however, is extremely sparse. The first identifiable plants with modern nucleated cells—the algae—go back at least 1.6 billion years, and we have an excellent record of early land plants beginning about 400 million years ago. We can see the first seed plants beginning some 350 million years ago, and of the flowering plants beginning about 120 million years ago. But the transitions between these major forms of plant life remain obscure.

Evolution, particularly of plants, is therefore more like a staircase than a handicap-access ramp. We have abundant information from the flat area of the steps, but are left to speculate about how plants made each leap to the next step. This pattern, in which major new forms of life seem to appear suddenly in the fossil record, can be considered a large-scale form of "punctuated equilibrium," a phrase coined by Niles Eldredge and Stephen Jay Gould in 1972 for the origin of new species.

Various reasons for the gaps leading up to new forms of life have been proposed. Intermediate forms may have lived in areas, such as dry uplands (proposed for the ancestral flowering plants), where fossilization is unlikely. In addition, transitional forms in rapidly evolving lineages may have been few in number and limited in their geographic distribution, so even if some were fossilized, it is less likely that they will be found. In contrast, once a new form of plant life established itself, it rapidly became quite abundant and colonized vast territories. These are the traditional explanations based on the intrinsic bias of the fossil record, but could there be something else, a more dramatic process behind some of the sudden stops and starts in the evolutionary record?

Darwin thought of evolution as proceeding gradually through the steady, step-wise process of natural selection, but in recent decades, genetic processes have been discovered that can create new forms of life more quickly. Several of these involve combining the resources of two or more very different organisms.

There is much evidence, for example, that ancient bacteria acquired new metabolic pathways through the promiscuous sharing of DNA. This came about through the bacterial habit of scavenging DNA from dead bacteria and incorporating it into their own genome. This can happen between species that are only distantly related. This is the major evolutionary process, unknown to Darwin, called horizontal gene transfer. It appears that parts of the photosynthetic machinery evolved separately in unrelated lineages of bacteria and were brought together in the first plantlike organisms through this ancient form of hybridization. This process, incidentally, is exploited by modern genetic engineers who manipulate bacteria to shuttle DNA from one organism to another.

We know also that the first algae came about through an intimate symbiotic association between two very different kinds of cells, an ancient predatory cell and a photosynthetic bacterium. Shape-shifters, such as amoebas, engulf their food by literally surrounding it with extensions of their body. One such ancient organism engulfed a cyanobacterium that was not digested but instead lived on and evolved into the first chloroplast. There is mystery, however, about what that first alga was like, what exactly its shape-shifting ancestor was, and how it subsequently gave rise to two rather different modern groups of algae: the reds and the greens.

Another sort of genetic shift may have occurred when the early land plants split into two very different lineages: the bryophytes (mosses, liverworts, and hornworts) and the vascular plants (everything else). Plants have two different phases to their life cycle: one that produces gametes and another that produces spores. The two phases are connected to one another, at least part of the time, but are quite different in structure. It is possible that in some early land plants, mutations in regulatory genes allowed whole sets of genes to turn on in parts of the life cycle where they were normally suppressed. Such an event may have led to a spore-producing phase in vascular plants, with growth capabilities that were previously available only in the gamete-producing phase.

This book will trace what is known about the history of plant life—how familiar features of modern plants gradually, or sometimes not so gradually, emerged over many millions of years of evolution. Concerning the breakthrough events of plant history, we naturally ask all the questions of a nosy reporter: who, what, when, why, how, and where. We still don't have all the answers, but we put together what we have into hypothetical scenarios—possible solutions to the various abominable mysteries—and test them with every fact that comes along. This is evolutionary biology.

For the purposes of this story, "plant life" is defined broadly to encompass all photosynthetic organisms, though the term "plant" is usually reserved today for the multicellular photosynthetic organisms that live on land. To tell the story of the plant way of life, however, one must begin with the origin of photosynthesis itself and explore the events that shaped the photosynthetic cells that eventually were able to survive on land.

It is hoped that by taking this journey, you will acquire a deeper understanding of how plants have coped over the ages with the challenges of life, as well as an enhanced appreciation for these diverse organisms upon which we so vitally depend for food, oxygen, medicine, building materials, aesthetic pleasure, and so much more.

Plant Life

Figure 1.1 Modern stromatolites like these in Shark Bay, Western Australia, are composed of alternating layers of organic remains and sediment. A mat of living organisms on the top continues to add to the structure. Stromatolites as old as 3.5 billion years appear to have been built in the same way. Photograph by Paul Harrison, posted on Wikimedia Commons, licensed by Creative Commons.

1

The Origin of Photosynthesis

In a few shallow, highly saline lagoons along the west coast of Australia, peculiar knobby pillars of rock called stromatolites (Fig. 1.1) stand like the disarrayed remnants of a terracotta army, eroded and distorted beyond recognition. These monoliths were not carved by some ancient civilization, however, but built up in fine layers by microscopic living organisms.

The builders of the stromatolites are primarily photosynthetic bacteria called cyanobacteria. Formerly called "blue-green algae," these abundant organisms lack the nuclei and other internal organelles of true algae and other higher forms of life (eukaryotes). Many cyanobacteria live as free-floating plankton or tangles of filaments (Fig. 1.2) and account for 20–30% of the photosynthesis occurring in open waters (Pisciotta et al. 2010; Waterbury et al. 1979). Stromatolite-building species, on the other hand, secrete a mucilaginous glue and stick together in thin mats on rock surfaces.

Microbial mats on the tops of the stromatolites extend them slowly upwards. Particles of sediment and lime precipitated from the water get trapped in the sticky matrix, and periodically bury the living microbes. The resourceful cyanobacteria in those instances migrate to the top of the sediment and begin a new mat. This results in a fine structure of alternating light and dark bands.[1] Stromatolites are thus, like coral reefs, built by the living organisms that inhabit them.

Stromatolites turn out to be one of our most important clues as to the origin of photosynthesis: the extraordinary light-driven process that creates carbohydrate and releases oxygen. Stromatolites have been around for about 3.5 million years, and are, in fact, among the earliest signs of any kind of life on this planet. They are abundant throughout much of the geological record, but rather scarce in rocks less than 500 million years of age. It is believed that during the great Cambrian Explosion (ca. 500–600 million years ago), when many new kinds of animals appeared, stromatolites came under attack by voracious grazing animals equipped with hard, scraping mouth parts. After that, they survived only in restricted sites too salty for such animals.

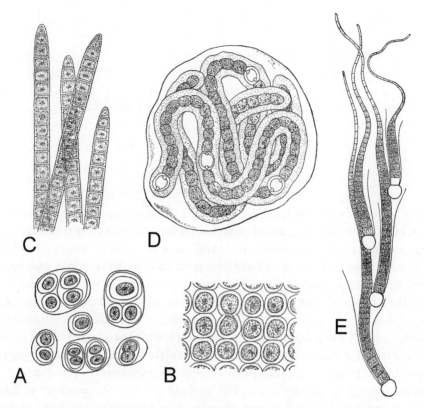

Figure 1.2 Modern cyanobacteria occur in many configurations, from unicellular plankton to filamentous or surface colonies: *Gloethece* (A) and *Merismopedia* (B) form flat colonies embedded in a mucilaginous sheath; *Oscillatoria* (C) consists of cells connected in long filaments; *Nostoc* (D) consists of filaments embedded in a mucilaginous sheath, with the round, clear cells specialized for nitrogen-fixation; *Rivularia* (E) colonies consist of tapered, hairlike filaments. Drawings by Coulter et al. 1910.

The presence of cellular remains in the most ancient stromatolites is debatable, but because they are similar in structure to modern stromatolites, it is generally accepted that they too were formed by photosynthetic bacteria, though possibly by earlier forms that did not release oxygen as a byproduct. Apparent fossilized cells have however been found in slightly different types of rocks, the Warrawoona and Apex cherts of Australia, which also date back to 3.5 billion years ago. J. W. Schopf and his associates identified these as cyanobacteria, based on their filamentous structure (Fig. 1.3) (see Schopf 2006 for discussion), and recent evidence seems to confirm that they were at least some kind of living organism (Derenne et al. 2008). If Schopf has identified them correctly, that would make the cyanobacteria the oldest known fossil organisms. They would in fact be nearly as old as the last universal common ancestor of life (LUCA), which is estimated to have appeared some 3.5–3.8 billion years ago (Boyd and Peters 2013).

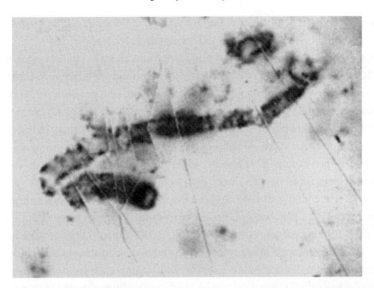

Figure 1.3 The apparent remains of cyanobacteria from 3.5 billion-year-old rocks in Australia. Photograph from Awramik 1992.

Though there has been controversy about their earliest appearance, there is little doubt that cyanobacteria were present by at least 2.8–2.5 billion years ago, which makes them still quite ancient and still the earliest form of life to leave a significant fossil record. In rocks of this later time period, their cellular remains are clear and remarkably like some modern species. Cyanobacteria left an important chemical clue as well: 2-methylhopane, a unique compound found almost exclusively in cyanobacterial cell walls (Summons et al. 1999, and Rasmussen et al. 2008). Though some other kinds of bacteria produce this compound, they exist in much more limited habitats, and are not likely to have ever been abundant enough to leave a trace in the rocks.

The early photosynthesizers must have gradually built up enormous populations, for the oxygen they produced eventually transformed the vast oceans of our planet, and then the very atmosphere itself. The scarcity of oxidized[2] ("rusted") minerals in the Earth's oldest rocks (for the sake of discussion, older than 3 billion years), indicates that there was very little free oxygen in the atmosphere at that time, so signs of significant oxidation in later rocks is evidence of plant life. Iron is particularly abundant on Earth, and it is quite prone to rusting. In the ancient seas there was a steady supply of iron bubbling up from underwater volcanic fissures and from eroding surface rocks. In its unoxidized state, iron is soluble in water, but when it oxidizes ("rusts") it forms insoluble molecules of hematite or magnetite, which sink to the bottom of the sea. When oxygen became available in comparably huge quantities there were spectacular depositions of iron. This resulted in distinctive and extensive rock layers known as the banded iron formations. The "rusting of the earth," as it was called by Schopf, is the source of most of the iron ore that is being ravenously consumed by modern civilization.

There is some evidence of limited iron formations about 3.5 billion years ago, but they did not become truly massive until the mid-2-billions. This suggests that oxygen buildup may have occurred sporadically and slowly at first, but became overwhelming between 2.7 and 2.5 billion years ago. Oxygen makes up 21% of our current atmosphere, but even more photosynthetically generated oxygen is tied up in banded iron formations and other oxidized minerals (Schlesinger 1991). The formation of iron deposits declined rapidly after about 2 billion years ago, as the supply of dissolved iron was depleted, and oxygen then began to build up in the atmosphere. The massive production of oxygen that triggered the banded iron formations suggests that oxygen-producing cyanobacteria first appeared much earlier, somewhere between 3.5 and 2.9 billion years ago (see Des Marais 2000, Schopf 2006, and Nisbet et al. 2007), in agreement with other evidence. The transformed atmosphere made it possible for more complex organisms to evolve, leading to eukaryotic cells (larger cells with nuclei and other internal organelles) and eventually multicellular animals.

Cyanobacteria are the only bacteria that conduct photosynthesis in virtually the same way as all higher plants, that is, with a release of oxygen gas as a byproduct. So they can be considered the first "plant life." They alone filled the ecological role of plants in the ancient ecosystem, pumping food energy into the world's food chain for roughly 2 billion years before other forms of photosynthetic organisms became abundant. If cyanobacteria were in fact present 3.5 billion years ago or even somewhat later, it would also mean that photosynthesis, as we know it today, evolved very early in the history of life.

From this ecological perspective, it would not be surprising if plants were the Earth's first inhabitants. Animals, fungi, and most microbes are dependent not only on the sugar produced by photosynthesis, but also on the oxygen it releases. These other organisms are heterotrophs, which obtain fuel energy from the food they consume and burn it through aerobic respiration. This process essentially reverses photosynthesis and balances it out in the great circle of life. I should hasten to add that plants themselves are functionally heterotrophs at night time, as are roots and other non-photosynthetic organs. In any case, it would seem that animallike organisms could not exist until there were plants. We will see shortly, however, that there are simpler metabolic ways of life that most likely preceded those of both plants and animals.

Photosynthesis:

carbon dioxide + water   **sugar + oxygen gas**
$$6CO_2 + 6H_2O + sunlight \rightarrow C_6H_{12}O_6 + 6O_2$$

Aerobic respiration:

sugar + oxygen gas   **carbon dioxide + water**
$$C_6H_{12}O_6 + 6O_2 \rightarrow 6CO_2 + 6H_2O + \text{cellular energy}$$

Photosynthesis is indeed a marvelous process: elegant, efficient, and virtually unlimited in its potential. With a practically inexhaustible supply of energy from the sun, and requiring only carbon dioxide and water as starting ingredients, photosynthesis provides a bountiful supply of both food and oxygen for a world teeming with complex life forms.

Where did photosynthesis come from?

Photosynthesis is one of the most complex of all biological processes, and consists of two very different but closely linked processes. In the "light reactions," solar energy drives the flow of energized electrons through tiny electrical circuits that power the production of the two major energy transfer molecules: ATP and NADPH. Those molecules in turn fuel the "dark reactions" (or better called the light-independent reactions), an enzyme based chain of reactions in which carbon is attached to an organic backbone and energy-rich glucose is created. Despite their traditional names, the light and dark processes occur at the same time, with sugar production more-or-less keeping pace with the supply of ATP and NADPH molecules from the light reactions.

Both the animal and plant ways of life are complex processes, and must have evolved in small bits over long periods of time. The light reactions in particular involve dozens of pigments and specialized electron-handling complexes lined up in precise arrays on cellular membranes, and the sugar-manufacturing process requires a dozen specialized enzymes. If life originated from organic chemicals formed by nonliving processes, it must at first have been extremely simple. Unfortunately, the fossil record does not help us out much with this earliest phase of life.

Fossilization is not an equal-opportunity process. Most organisms decay and completely disappear after they die, and this is probably what happened to virtually all prephotosynthetic life. In order to leave its calling card in the rocks, a particular type of organism must be quite abundant, live somewhere where fossilization (burial with minimal decay) is taking place, and have body parts, or unique chemical products, durable enough to remain in the rocks for millions or even billions of years. Once perfected, photosynthesis, and the cyanobacteria that possessed it, were so successful that they dominated the planet for 2 billion years and left abundant remains in the rocks. These organisms had cell walls that were durable enough to leave outlines in the sediments, at least some of the time, and they also occurred in dense local colonies (e.g., stromatolites), which made it much more likely that their remains would be permanently preserved in the rocks.

So if there were simpler forms of life before cyanobacteria, and even if there were animallike (heterotrophic) or other forms of life that coexisted with cyanobacteria during their first billion years, there is virtually no fossilized record of their existence. In tracing the early history of life, biologists turn to other sorts of evidence: the genetics and biochemistry of living descendants of bacteria that lived long ago. Remarkably, some very ancient processes for obtaining energy are still practiced, and some have been incorporated into more modern processes like photosynthesis. We know this because both the genetic instructions and the details of the biochemistry are basically the same. Most surprising is the revelation that modern, oxygen-generating photosynthesis contains within its complex machinery pieces that came from at least three more ancient kinds of organisms.

Chemosynthesis and the Calvin cycle

The ability to make sugar most certainly preceded the ability to capture energy from light. Sugar, or even simpler organic molecules, can be made only if some external source of energy is available to force inorganic carbon compounds to join together chemically. In photosynthesis, light energy drives a sequence of energy conversions that ultimately joins inorganic carbon dioxide into energy-rich sugar molecules. In organisms called chemosynthesizers[3], energy from inorganic minerals is employed instead.

Inorganic compounds with retrievable energy are relatively rare on the Earth today. Most minerals in the Earth's crust have been oxidized. But substances such as hydrogen gas or hydrogen sulfide (H_2S), bubbling up from volcanic vents under the ocean or from terrestrial hot springs, provide the energy base for lively assemblages of microorganisms. Some chemosynthetic processes are quite simple and may have been used by the earliest cells before either photosynthesis or animal-like heterotrophs evolved.

Today, whole ecosystems subsist around volcanic vents in the inky black depths of the oceans. Heterotrophic microbes consume chemosynthetic bacteria and in turn are preyed upon by sequentially larger animals. Some chemosynthesizers even live within the tissues of animals such as giant tube worms and exchange energy and nutrients with them through direct symbiosis.

To make a long story short, it is now widely believed that the first living organisms were chemosynthesizers living around ancient volcanic vents, and it was they that invented the process of making sugar. There are many competing theories as to the origins of life itself, but only organisms with an external source of energy, such as chemosynthesizers and photosynthesizers, manufacture their own carbohydrate fuel. Though there are variations on exactly how sugar is synthesized among bacteria, the process became standardized in advanced organisms as the Calvin cycle (or to give credit to Calvin's oft forgotten colleagues, the Calvin-Benson-Bassham cycle), and in that form it was adopted by the first photosynthetic organisms.

The Calvin cycle is a step-by-step process for adding carbon dioxide to a pre-existing organic molecule and converting it to energy-rich glucose. Each step is facilitated by a specific enzyme, and the critical step of attaching a molecule of carbon dioxide is moderated by the most abundant enzyme in the world: ribulose-1,5 bisphosphate carboxylase, known as rubisco, for short. The Calvin cycle is called a cycle because some of the product is recycled back into the starting molecule to which carbon dioxide can be attached (Fig. 1.4). There is no biological process for attaching carbon dioxide molecules directly to one another, hence this roundabout process.

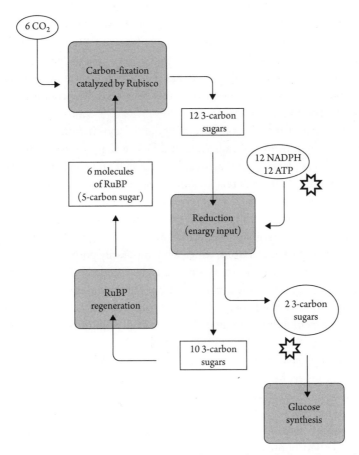

Figure 1.4 The generalized flow of materials and energy through the Calvin cycle. It is a cycle in the sense that one of the end products (RuBP) is also the primary starting reagent. Carbon dioxide combines with the 5-carbon sugar RuBP, which breaks down into more stable 3-carbon intermediaries. The intermediaries are then reduced (energy added from ATP and NADPH), after which a small percentage forms into glucose, and the rest are rearranged to form RuDP again. For every six molecules of carbon dioxide (CO_2) entering the cycle, one molecule of glucose is generated. Sun symbols represent energy input and output.

Energy is required to drive the Calvin cycle and create the high-energy bonds of glucose. That energy is provided in chemosynthesizers through the oxidation of energy-rich inorganic molecules, but in plantlike organisms, it comes through the capture of sunlight.

The light reactions

As early chemosynthesizers proliferated in the depths, some undoubtedly migrated to more brightly lit waters, perhaps in sulfurous surface pools like those found in the hot springs of Yellowstone National Park. Here, life encountered another threat: ultraviolet radiation, which was far more intense 3.5 billion years ago than it is today. The ozone (O_3) layer that gives us partial protection against ultraviolet (UV) rays today didn't exist then, because there was no oxygen (O_2) in the atmosphere from which it could form. It is likely that this promoted the evolution of protective pigments.

Some modern cyanobacteria have a pigment (scytonemin) in their sheaths that protects them from UV radiation, and before that, the precursors of chlorophyll may have played a similar role. Chlorophyll and the hemoglobin in our blood are actually chemically related and appear to have evolved from a common ancestral pigment, even though they have taken on radically different functions.

By definition, pigments are molecules that absorb particular wavelengths of light energy and reflect others, giving them color. The light energy absorbed by pigments is taken on by particular electrons, which are temporarily pushed to higher orbitals around their nuclei. Such electrons are said to be "excited." If there is nothing to hold them in that position, the excited electrons fall back immediately to their normal orbitals. The absorbed energy is then released in a less intense form, typically as heat or a burst of fluorescent light. A test tube full of purified chlorophyll in fact does exactly that, accomplishing nothing useful at all.

In the living cells of cyanobacteria and the chloroplasts of plants, however, excited electrons are held up, and their energy is released in a controlled, step-wise manner. Some of that energy is used to make stable, energy-rich molecules that can be used by living cells.

Light energy is captured and converted to chemical energy in organized complexes called photosystems, which reside in precise patterns on the inner membranes of cyanobacteria and chloroplasts, and function much like electrical circuits. The electrons are excited in clusters of chlorophylls and other light-sensitive pigments called antenna complexes. From there, the electrons pass down a series of molecules called an electron transport chain, which controls the gradual release of their excess energy and returns them back to the antenna complexes.

We might presume that the cyclic electron flow around simple circuits such as these began as a way to safely dissipate the energy absorbed from high energy radiation, but the potential to put those circuits to work probably did not stay idle for long. The "downhill" flow of electrons along the electron transport chain is comparable to running water, which can be harnessed to grind grain or spin a turbine to generate electricity. The breakthrough that changed the world came when light-driven electron flow was linked to existing machinery for making the stable energy compound adenosine triphosphate (ATP).

Making ATP actually consists of adding a phosphate unit to simpler molecules called adenosine diphosphate (ADP), creating a high-energy bond in the process. The technical name for this process is cyclic photophosphorylation. ATP is the primary energy courier in all living cells. It provides the energy input for the vast majority of metabolic reactions in living cells, for processes such as nerve impulse transmission and muscle contraction, as well as for the manufacture of sugar in the Calvin cycle.

Protons and ATP

The flow of electrons does not directly create ATP. Instead, the electron flow is coupled to an even older process involving the flow of protons (positively charged hydrogen ions). The energy of the excited electrons makes a big drop as they pass through a part of the circuit called the cytochrome complex. Here, the electrons are drawn toward the inside of the chloroplast membrane and they drag with them positively charged protons (hydrogen ions). This is called a proton pump. In this process, energy is transferred from the electrons to the protons.

With the energy absorbed from light now depleted, the electrons continue around the circuit back to the chlorophyll molecules, but the protons are trapped in specialized internal chambers called thylakoids. This results in concentrated, battery-like pools of protons—a form of potential energy. The tendency of concentrated charged particles is to return to less concentrated regions of the cell (balancing out both the electrical charges and the particle concentration). This creates a proton motive force. The only escape from the thylakoids is through channels in another molecular complex in the cell membrane: ATP synthase (Fig. 1.5). As the protons flow through this complex, like water through an old-fashioned grist mill, they drive the synthesis of ATP.

The proton motive force is widely, if not universally, used by living organisms, primarily to generate ATP. It occurs not only in photosynthesis, but also in cellular respiration processes in mitochondria and bacteria. The coupling of this process with the capture of light energy marked the beginning of plant life.

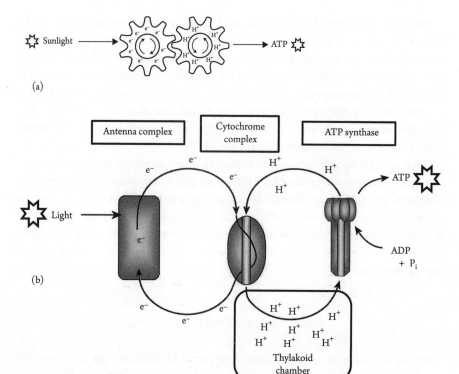

Figure 1.5 The energy of sunlight drives the production of ATP in a two-step process analogous to the turning of gears (inset). Electrons energized by chlorophyll flow through the cytochrome complex, which drives protons into a chamber, then return to the chlorophyll; the hydrogen ions flow back out of the chamber through the ATP-synthase complex, which drives the manufacture of ATP molecules.

In the simplest forms of photosynthesis, therefore, sunlight drives the circulation of electrons, electrons drive the circulation of protons, and the flow of protons then drives the synthesis of ATP. This, however, is just the beginning.

NADPH—a more powerful energy courier

A variety of bacteria, which we will call sulfur bacteria to keep things simple, employ an ancient form of chlorophyll called bacteriochlorophyll and a photosystem modified to synthesize not only ATP, but also a more powerful energy molecule called NADPH. Both ATP and NADPH are needed by the modern Calvin cycle. The modified system is non-cyclic—the excited electrons do not return to chlorophyll but become incorporated directly into NADPH, imparting their energy to that molecule.

The Origin of Photosynthesis

Since the excited electrons do not return to the bacteriochlorophyll molecules, electrons must be pulled from somewhere else to replace them if the process is to continue. That source is most often hydrogen sulfide, which releases electrons fairly easily, leaving elemental sulfur and hydrogen ions as byproducts. So the various kinds of sulfur bacteria can exist only in restricted habitats, such as hot springs, where sulfurous compounds are available.

The ability to make NADPH was a huge advance, but if it had stayed at that, photosynthesis would have supported only modest and limited food chains—and there would be no oxygen in the atmosphere to support higher forms of life. Another huge advance came when a photosystem was modified to use ordinary water instead of hydrogen sulfide as the electron source. This not only opened

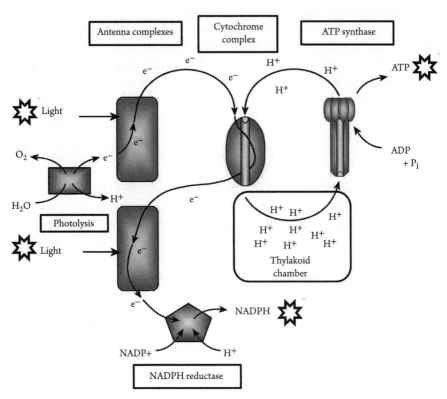

Figure 1.6 In non-cyclic electron flow, excited electrons flow through the cytochrome complex, but instead of returning to the chlorophyll molecules, they move to a second photosystem, where they are re-excited by sunlight and ultimately become incorporated into NADPH molecules. Electrons removed from the first photosystem are replaced by electrons taken from water molecules in the process of photolysis.

up vast new habitat for incipient plant life, but also resulted in the release of the all-important byproduct, oxygen gas (O_2).

It is much tougher pulling electrons out of water than out of hydrogen sulfide, however, so a stronger electron-grabbing force is required. The "suction" of an electron transport chain—the strength with which it can pull electrons from other substances—is technically reduction (or redox) potential. Hydrogen sulfide (H_2S) and water (H_2O) have an obvious structural similarity, but oxygen has a higher redox potential than sulfur.

The redox potential sufficient for pulling electrons out of water was achieved by the ancestors of cyanobacteria by hooking two photosystems together in tandem (Fig. 1.6), along with the evolution of a more advanced central pigment called chlorophyll a. Think of the flow of excited electrons leaving bacteriochlorophyll as creating a vacuum that sucks electrons out of hydrogen sulfide. Then imagine hooking two such vacuum cleaners together, creating enough of a pull to remove electrons from water.

The starting ingredients for this new form of photosynthesis—water and carbon dioxide—were extremely abundant, and so photosynthesis became virtually unlimited in its potential productivity. The cyanobacteria thrived, leaving an abundant fossil record for the past 3 billion years and a continued massive presence today. The sulfur users were and still are strictly limited to rare sulfurous habitats and have left few fossils.

Horizontal gene transfer

This brief and greatly simplified account of photosynthesis is provided in order to answer the question of how it came about in the first place. We are accustomed to thinking of evolution as proceeding through the painstaking process of random mutation, followed by many generations of natural selection. We now know, however, that more dramatic kinds of change can occur through the process of horizontal gene transfer, which is remarkably common among bacteria.

The familiar process of hybridization in plants and animals can normally occur only between closely related species, but bacteria are much more liberal about accepting foreign DNA. Many of them routinely absorb DNA from dead bacteria and incorporate it into their own genome. It can then be expressed along with the original DNA of the cell.

This process of transferring genes horizontally can occur between distantly related species, resulting in the combination of very different metabolic capabilities. ("Vertical gene transfer" would be the more routine inheritance of genes from one's parents.) We can presume that this often leads to dysfunction, but occasionally, the acquired DNA works with existing DNA to create new biochemical pathways. Horizontal gene transfer, incidentally, is what is exploited when bacteria are used to transfer genes from one organism to another in genetic engineering.

Such transfers appear to have happened at least twice during the evolution of photosynthesis: once to combine the Calvin cycle with a simple photosystem, and again when the second photosystem was added (see Raymond et al. 2002). The two photosystems, named photosystem I and photosystem II, were apparently acquired from different ancestral bacteria, which used them in solo mode to capture light energy. Photosystem I is similar to the single photosystem found in green sulfur bacteria and heliobacteria, while photosystem II is similar to the one found in green gliding bacteria and purple non-sulfur bacteria. We believe the two were combined through the process of horizontal gene transfer in the ancestors of cyanobacteria.

The two photosystems are similar enough that they most likely had a common ancestor in an even more ancient type of bacterium. In photosystem I, the beginning of the electron transport chain includes iron-sulfur clusters, while in photosystem II, those clusters have been replaced by other molecular complexes. Iron-sulfur particles are believed to have been important catalysts for the first chemosynthetic life around volcanic vents, so most likely, photosystem I is the older one. Archaic heliobacteria use a minimal cyclic process to make ATP but are not capable of making carbohydrate like other photosynthetic bacteria or plants. These photoheterotrophs need to consume organic compounds for use in cellular construction. It is possible that bacteria similar to these preceded the joining of light reactions to the Calvin cycle that led to the first bacterial photosynthesis (Asao and Madigan 2010).

An alternate hypothesis is that the first photosystem evolved piecemeal in an ancient chemosynthesizer that was already using a form of the Calvin cycle. In this hypothesis, bacteriochlorophll originally served the purpose of sensing and moving toward sources of heat, where the minerals needed for fuel would be most abundant. (See Niklas 1997, for a good discussion of this whole topic.)

Cyanobacteria and the nitrogen crisis

The three elements needed for making carbohydrates—carbon, oxygen and hydrogen—were virtually unlimited and freely accessible to the early plants, so it would seem that there was no obstacle to their productivity. But there is one more element required in large amounts to make many of the other molecules needed for life: nitrogen. Nitrogen is an essential ingredient in proteins, DNA, and many other vital molecules, yet it is often very limited, at least in forms that can be used by organisms. Farmers don't need to fertilize with carbon, oxygen, or hydrogen, but they must constantly replenish the nitrogen in the soil.

This element is extremely abundant in the atmosphere as nitrogen gas (N_2), but ironically relatively few living organisms can use it in that form. The triple bonds that hold the two nitrogen atoms together are difficult to break, and energy is required to do so. Cyanobacteria are one of the major groups of organisms that have the enzyme (nitrogenase) necessary to break that bond and convert atmospheric nitrogen into ammonia. It was the final piece of metabolic machinery they needed to become the most abundant organisms on the planet, and to dominate, virtually alone, for over a billion years.

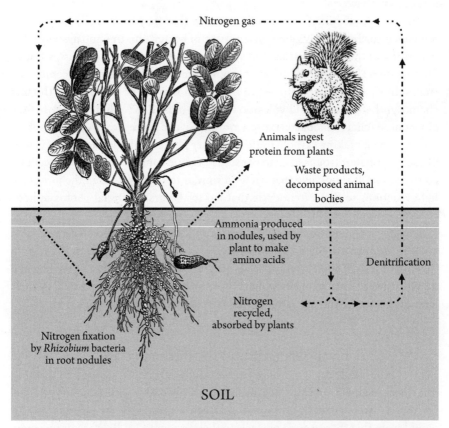

Figure 1.7 Nitrogen gas (N_2) is abundant in the Earth's atmosphere, but in a stable form unavailable to most living organisms. Certain bacteria convert the nitrogen to ammonia, which can be absorbed by plants or converted to other forms by various soil bacteria. Animals consume plants, and decomposers consume dead animals, animal waste, and dead plant material. This returns nitrogenous waste to the soil where some of it is recycled by plants. More of it is broken down by bacteria in the soil, eventually returning it to the atmosphere as N_2. Redrawn from several classical sources, including Brown 1935.

Plants can absorb and use nitrogen if it is either oxidized to nitrate (NO_3^-) or reduced to ammonia (NH_3), and animals can obtain it only by eating plants or by eating animals that have eaten plants. Ammonia and nitrate that find their way into the soil, from mineral sources or through decomposition of dead plants and animals, disappear quickly. Plants reabsorb some of it, but a host of microorganisms compete for it as well, and when they're done with it, it goes back into the atmosphere as nitrogen gas (N_2). Yet life continues in great abundance. How, then, is nitrogen made available to the living world?

While a significant quantity of nitrogen gas is oxidized to nitrate by lightning in the air, thus becoming available to plants, at least half of the nitrogen needed by the living world is provided by bacteria through biological nitrogen fixation. It is still uncertain where or when this important process evolved. It has variously been proposed to have been present in the common ancestor of life (LUCA), or to have evolved in one group, and then been passed on to others through horizontal gene transfer (Raymond et al. 2004). It is widespread, in several different forms, among bacteria, and in some clades of archaea,[4] but never in eukaryotes. It is common in cyanobacteria and closely related clades, but absent in some of the most ancient lineages of bacteria. The most ancient branch of bacteria is considered to be the Aquificales, of which some practice nitrogen fixation and others do not. In those that do, there appears to have been a later retrofitting (Boyd et al. 2013). The origin of nitrogen fixation thus remains obscure, and its widespread occurrence is most likely due to horizontal gene transfer.

It may have required a specific "nitrogen crisis" to jumpstart the process—a time when increasing abundance of life forms, buildup of oxygen, and decreased availability from rocks made nitrogen a limiting factor for growth. This may have been as late as 2.2 to 1.5 billion years ago (see Boyd et al. 2013, for a good review of this topic). If the latter is true, cyanobacteria acquired it relatively late in their two billion year reign as supreme photosynthesizers.

To make things more difficult, nitrogenases are highly sensitive to oxygen, and so nitrogen fixation must be shielded from exposure. Some nitrogen-fixing bacteria are strictly anaerobic, but for aerobic organisms, this is a problem. Cyanobacteria of course generate oxygen as part of photosynthesis. In some, nitrogen fixation is isolated in specialized nonphotosynthetic cells (see Fig. 1.2D). In others, nitrogen fixation takes place at night, when there is no photosynthesis. Nitrogen fixation in the terrestrial environment is carried out primarily by bacteria of the genus *Rhizobium*, which live symbiotically in special nodules on the roots of legumes (family Fabaceae) (Fig. 1.7).

As one more addendum, recent phylogenetic studies (see Hemp and Pace 2010) of the molecular complexes involved with oxygen metabolism in a wide variety of organisms indicate that the genes required for aerobic respiration were transferred from cyanobacteria to a number of different bacteria and archaea via horizontal gene transfer. So cyanobacteria ultimately gave us both sides of the great energy cycle, and possibly nitrogen fixation as well.

Figure 2.1 Elysia clarki is a sea slug, a relative of snails and garden slugs, that stores captured chloroplasts within its tissues and lives off of their productivity. Its story parallels an important phase in the evolution of plants. Photographed in the laboratory of Sydney K. Pierce, University of South Florida

2

Eukaryotic Plant Life

Have you ever imagined what it would be like if humans could photosynthesize like plants? Science fiction writers have speculated on it for decades, and with all the latest developments in genetic engineering, it is an idea that seems quite credible. But is it practical? Would it allow us to survive without food? Would we even want to live without food or spend every daylight hour sunbathing?

The idea is impractical in many ways. For one thing, we're not shaped right for photosynthesis. There are too many cells in our dark interior to be supported by a thin veneer of photosynthetic skin; photosynthesis could provide only a small fraction of our energy needs. And in obtaining that small fraction, the long hours of sunbathing would subject us to massive skin damage and the risk of cancer.

While the idea of photosynthetic humans may be far-fetched, there are, amazingly, some animals that have managed to retrofit themselves to live like plants, including the sea slugs pictured above (Fig. 2.1). They do so by salvaging chloroplasts from the algae they eat, and maintaining them within their tissues. Chloroplasts are the discrete cellular organelles that conduct photosynthesis in eukaryotic algae and land plants. The photosynthetic sea slugs have also evolved a flat, leaflike shape, which provides sufficient light-gathering surface area to supply the animal's needs for carbohydrate.

Most of these sea slugs must periodically replace the chloroplasts, but at least one species, *Elysia chlorotica*, can maintain a set of chloroplasts obtained in its youth and live off of them for rest of its short life (Rumpho et al. 2008). In this species, some of the genes required for chloroplast maintenance have even been transferred to the sea slug's nuclei through horizontal gene transfer, the first confirmed instance between multicellular eukaryotic organisms (see Pierce et al. 2003). This is a step toward taking control of the chloroplasts and making them one's own, but sea slugs will not be truly and permanently plantlike until they are able to replicate the chloroplasts and pass them on to their offspring.

As an isolated story, that is certainly interesting, but the startling fact of the matter is that the ancestors of plants, as well as a great many plant pretenders, got their start in virtually the same way.

Mutualism and kleptoplasty

So valuable is the potential of photosynthesis that many animals have found ways to incorporate it into their own survival strategy. One means to that end is to partner up with organisms that are already photosynthetic. A number of sponges, hydras, sea anemones, coral polyps, flatworms, and clams harbor single-celled algae within their tissues, and exchange nutrients for photosynthetic product. This is a mutually beneficial symbiosis known as mutualism, and without which most of these animals would not survive.

Many reef-building corals are dependent on internalized algae (dinoflagellates of the genus *Symbiodinium*) for the energy needed to deposit their calcium carbonate encasements. If these domesticated algae die off, the coral colonies and the entire reef system may go into decline. The animals are not photosynthetic themselves, but such intimate symbioses give them almost the same functionality.

In the terrestrial world, similar partnerships between fungi and photosynthetic microbes, known as lichens, are so convincing that they were long classified as a group of plants (Fig. 2.2). A lichen is an intimate, mutually dependent, and beneficial relationship in which the fungal matrix soaks up rainwater and minerals and makes these available to the algae. The algae (or in some cases cyanobacteria) repay the debt by passing some of their photosynthetic product back to the fungus. Lichens are among the most hardy of "organisms," surviving on tree bark and bare rocks that are only occasionally wetted by the rain.

The sea slug approach is more direct, almost brutal: the chloroplasts are literally stolen from other organisms. This is more like an organ transplant than symbiosis.

The habit of ingesting and exploiting chloroplasts from other organisms was dubbed kleptoplasty, ("stealing of chloroplasts") by Kerry Clark in the 1980s, and the stolen chloroplasts are often referred to as kleptoplasts. Sea slugs are unique in having acquired chloroplasts as multicellular animals, but over the past billion years or so, it is likely that hundreds of single-celled predators have independently acquired chloroplasts in a similar manner.

There exist today a number of unicellular organisms that routinely ingest and retain functioning chloroplasts within them, including some amoebas, heliozoans, foraminiferans, and a great many ciliates (*Paramecium* and its relatives). In some marine planktonic communities, more than 40% of the ciliates contain captured chloroplasts (Stoecker et al. 1987). These chloroplasts must be replaced periodically and do not pass on to the host cell's offspring. Those in the past that overcame that limitation became the permanently photosynthetic organisms we call algae.

There is one thing missing from this story: if organisms become photosynthetic by stealing chloroplasts, where did the first chloroplasts come from? Before

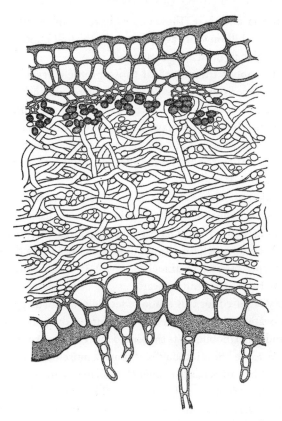

Figure 2.2 Lichens are a symbiotic association between a fungus and algal cells. The algae (spherical cells) provide photosynthetic product to the fungus, which provides water and minerals, as well as protection, for the algae. Drawing from Brown 1935.

there were eukaryotic algae, photosynthesis was exclusively practiced by bacteria. The very first chloroplasts then were domesticated cyanobacteria. This first event was not kleptoplasty, but a mutualistic symbiosis, for what was captured was a whole organism that was kept alive within the host cell. As permanent captives, the cyanobacteria exchanged nutrients and materials with their host cells and thrived within their protective environment. Eventually they lost the ability to live independently and became cellular organelles.

The first eukaryotes

The early cells capable of ingesting and domesticating other cells were larger and more flexible than their prokaryotic neighbors. They had distinct nuclei, mitochondria, and other internal organelles. These were eukaryotes: the ancestors of all animals, plants, fungi, and countless kinds of single-celled organisms like amoebas, paramecia, and algae. These larger, more complex cells began to appear probably around 2.5 billion years ago. They were the first hunters, and cyanobacteria were probably their most abundant prey.

We know little about the nature of those protoeukaryotic cells or exactly how they began to consume other cells. They were evidently descended from an archaean, a prokaryote similar to true bacteria in size, shape, and simplicity of structure, but differing in many details of their cell walls, chromosomes and metabolism. The ancestral eukaryote appears to be most closely related to modern archaeans called thermoacidophiles, organisms that thrive in water of extreme temperatures or acidity.

These protoeukaryotes were remarkable in several ways and quite unlike other prokaryotes. They were flexible, "naked" cells without walls, and capable to some degree of changing shape. They at least could form pockets at their surfaces that, like tiny mouths, could pinch off around food items and bring them inside. They were early protoypes of modern amoebas (Fig. 2.3), single-celled predators that move and feed in this shape-shifting way.

The flexibility required for this mode of feeding was made possible by a most remarkable innovation: the cytoskeleton—a complex, dynamic system of protein rods, tubules, and small mobile units that could move things around within the cell. The cytoskeleton, as we know it today, is as much a muscular system as it is a skeleton. It enables cells like amoebas to change their shapes or move forward by shifting cytoplasm from one part of the cell to another. This results in the familiar creeping movement that may have been the inspiration for the 1950s horror classic *The Blob*.

Equally important is the ability to move things around inside the cell. Small packages, or vesicles, containing food, waste products, or other materials are carried by tiny motor proteins that "walk" along tracks within the cytoskeletal matrix. Chromosomes are moved around in a similar way during cell division. The ability to move things around inside allows eukaryotic cells to get much larger than prokaryotes, which rely on diffusion for internal distribution. So somewhere in the mid-2-billions, large sophisticated cells had gotten into the business of

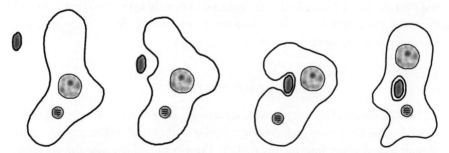

Figure 2.3 Amoeboid cells change shape by extending portions of the internal cytoskeleton. This allows the cell to move in a particular direction as well as to form pockets around food objects. Redrawn from Hartog 1906.

eating other cells. It is not hard to imagine then, that, like sea slugs, some of them domesticated cyanobacteria to become photosynthetic.

Endosymbiois

In the early 1970s a young biologist named Lynn Margulis achieved notoriety for reviving the "outrageous" theory that not only chloroplasts, but also mitochondria (aerobic energy-processing organelles) were originally free-living bacteria that came to dwell inside early eukaryotic cells (e.g., Margulis 1998). This scenario, first proposed by the Russian biologist Mereschkowski in 1905, had been derided, discarded, and largely forgotten by the scientific community until revived by Margulis and others of her generation.

Both mitochondria and chloroplasts retain vestiges of their former independent life, including their own bacterial loops of DNA and interior membranes. A few algae (glaucocystophytes) even retain vestiges of peptidoglycan, the unique wall material of bacteria, around their chloroplasts. The incorporation of whole cells, in this case ancient bacteria, into larger cells is called endosymbiosis. The endosymbiosis of ancient bacteria into larger cells, followed by their transformation into mitochondria and chloroplasts (Fig. 2.4), is now firmly accepted by the

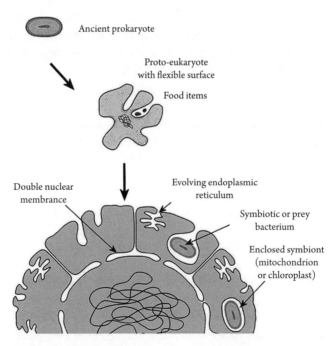

Figure 2.4 The evolution of the cytoskeleton and loss of the cell wall allowed the first eukaryotic cells to become larger and more flexible. Portions of the cell membrane extended inward to form vacuoles, endoplasmic reticulum, the nuclear envelope, and other internal organelles.

scientific community and presented as a matter of fact in freshman biology textbooks. Endosymbiosis leading to chloroplasts probably happened only once (possibly twice), while kleptoplasty, the theft of those chloroplasts from one eukaryote by another, happened many times.

Protists

The earliest eukaryotes, and numerous of their simple modern descendants, are by definition protists (also called *protoctists*). They are mostly aquatic organisms ranging from single-celled amoebas to multicellular algae ("seaweeds"). They represent a level of organization that is simpler than that of plants, animals, and fungi, and for many years were classified together as the kingdom Protista. Large seaweeds, such as the giant kelps, have three-dimensional tissues and distinct stem- and leaflike organs, and so are very much like the higher plants that live on the land. It is therefore hard to define precisely what a protist is, other than "eukaryotes that are not plants, animals, or fungi." True plants have a greater specialization of cells, tissues, and organs, and protect their young embryos within special chambers, something not found in any protist.

Modern phylogenetic analysis has revealed that the protists represent a wide and varied assortment of independent eukaryotic lineages. Animallike, plantlike, and fungus-like traits have evolved multiple times among them, resulting in unrelated organisms that happen to look or behave like each other (convergent evolution). The word "alga" refers to any of the great variety of photosynthetic protists, including multicellular seaweeds. Terrestrial plants, animals, and fungi—what we consider three of the major kingdoms of life—are relatively recent branches of some of these ancient protist lineages and are well defined. So the word protist represents a level of complexity, not a distinct taxonomic group.

Protists are often highly mobile, particularly those that have specialized as animallike predators ("protozoa"). The first protists were probably similar to amoebas, organisms that literally "surround" their food. Others evolved fixed, feeding pockets that funnel food down into food vacuoles. *Paramecium* is one of the most successful and widespread of these single-celled predators. Others, like *Vorticella* or *Diplodinium*, attach to rocks or underwater plants and just wait for food to come by, which they catch with tentacles surrounding their mouths.

Many protists move about by means of flagella, as do some bacteria, but with flagella of radically different structure and operation. Unlike the protein filaments that spin like propellers in bacteria, flagella in protists evolved from narrow extension of the cell itself, complete with cell membrane and cytoskeleton elements. These flagella contain a distinctive arrangement of long protein microtubules, consisting of a ring of double tubules and two single tubules in the center (Fig. 2.5). Such flagella are whiplike, waving back and forth as the microtubules

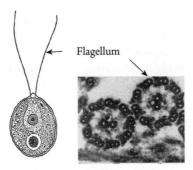

Figure 2.5 Two flagella of the green alga, *Chlamydomonas*, in cross section. This distinctive internal structure of nine double microtubules surrounding a pair of single tubules, is standard in the flagella of eukaryotes from algae to human sperm cells. Drawing from Haupt 1953; photograph courtesy the Dartmouth Electron Microscope Facility.

on either side alternately contract. While flagella are obviously of value to aggressive predatory cells, many single-celled algae also have them, as do reproductive cells of many multicellular seaweeds.

Lynn Margulis provided another idea, one that has not been so popular. According to her (see Margulis 1998), a bacterium with multiple simple flagella, something like a modern spirochaete, attached itself to an ancient eukaryotic cell, lost much of its own cellular structure, and evolved into the eukaryotic flagellum. In this model, the spirochaete contributed to the evolution of the cytoskeleton by partially moving inside the cell. Some modern protists, such as the *Mixotricha paradoxa* that live in the guts of termites, do in fact have spirochaetes attached to their cell surface, and they may help them move around. The idea that eukaryotic flagella evolved from such a symbiosis has always been highly controversial, and there remains little hard evidence for it.

Eukaryotes did not replace prokaryotes, but instead created vast new habitats for them. Billions of individual bacteria, representing hundreds of species, for example, inhabit each human body. But these are largely invisible. The world as we perceive it is made up of the much larger eukaryotes.

The major kinds of algae

Among living algae, we can recognize three general classes of chloroplasts: "red," "green," and "brown," to keep things simple. Each has a distinctive mix of light-gathering pigments and other unique structural features (see Graham and Wilcox 2000 for an excellent general reference.) Correspondingly, the three largest groups of algae are the red, green, and brown algae, each with the appropriate type of chloroplast. Brown and green chloroplasts have also found their way into some unrelated groups, as you might guess, through kleptoplasty.

The reddish hue of most red algae is due primarily to a specialized pigment called phycoerythrin, which absorbs mainly the blue and green parts of the visible solar spectrum, reflecting back red wavelengths. Blue light penetrates deeper into the water than the rest of the spectrum, so red algae are adapted to survive in these lower depths.

The red algae are a large, diverse, and highly successful group of organisms. Most are multicellular seaweeds (Fig. 2.6), and most are also marine. Many red algae consist of interwoven filaments of cells, while others have true three-dimensional cell division. Most can be recognized by a distinctive plug structure (pit plug) connecting one cell to another. Some are encrusted with calcium carbonate crystals that give them a rigid, coralline constitution. Many red algae, in fact contribute to the buildup of coral reefs.

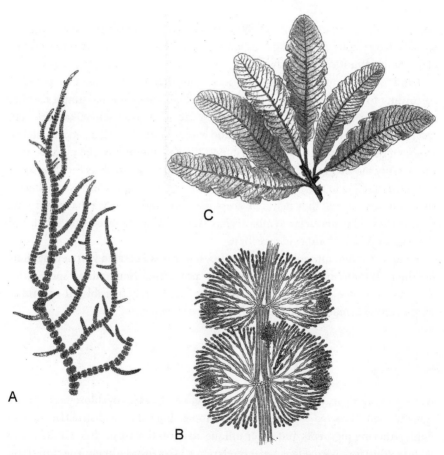

Figure 2.6 Red algae mostly take the form of macroscopic, multicellular seaweeds, and are incredibly varied in form. Some such as *Batrachospermum moniliforme* (A, B) are based on complex arrangements of filaments, while others such as *Delessaria sanguinea* (C) have leaflike structures with two-dimensional tissue formation. Drawing from Oltmanns 1905, attributed to Sirodot.

Green chloroplasts contain chlorophylls a and b along with certain carotenoids (pigments that give orange color to carrots and leaves in the fall) adapted to absorb the broader portion of the light spectrum, including both blue and red wavelengths, that is present in shallower water and full daylight. Green algae living in freshwater habitats were thus well poised to move onto land. We will devote the latter part of this chapter to this important group of organisms.

The odd brown coloration of brown algae (due to the carotenoid pigment fucoxanthin) enhances absorption of green light, the part of the spectrum generally reflected away by other pigments, perhaps giving them a competitive advantage at moderate to shallow depths. "Brown" chloroplasts are found not only in the brown algae proper, but also in diatoms, some dinoflagellates, and a few additional groups.

Brown algae are mostly complex, multicellular, seaweeds (Fig. 2.7), the largest being the giant kelps that form underwater forests along the Pacific coast of the United States. Diatoms are a highly eccentric group of photosynthetic protists that build ornamented, overlapping glass shells around themselves (Fig. 2.8), and are so abundant that they often leave massive deposits (diatomaceous earth) that can be mined and used as an abrasive.

Two groups of algae that appear to be relatively recent in origin are single-celled organisms with relatives that are still predatory: the euglenids (Fig. 2.9A), with green chloroplasts, and the dinoflagellates (Fig. 2.9B), which may have either brown or green chloroplasts. Both of these groups have originated through kleptoplasty.

Primary and secondary endosymbiosis among algae

The red and green algae appear to have descended directly from the first endosymbiosis between a eukaryotic predator and a cyanobacterium. Much DNA evidence suggests that this event—called primary endosymbiosis—occurred just once. Algae that have brown chloroplasts, and a few that have green chloroplasts, all appear to have arisen through kleptoplasty followed by permanent integration of the captured chloroplasts—a two-step process known as secondary endosymbiosis.

Secondary endosymbiosis leaves a tell-tale trail of extra membranes surrounding the chloroplast, for each time that an ingested chloroplast became a permanent resident, the membrane from the food vacuole that formed around it typically remained as a permanent envelope. The chloroplasts in red and green algae have two membranes around them, one that belonged to the original cyanobacterium, and one that came from the food vacuole of the ancient eukaryote that ingested that cyanobacterium.

Dinoflagellates, euglenas, diatoms, and brown algae have three membranes around their chloroplasts, having added another food vacuole membrane when

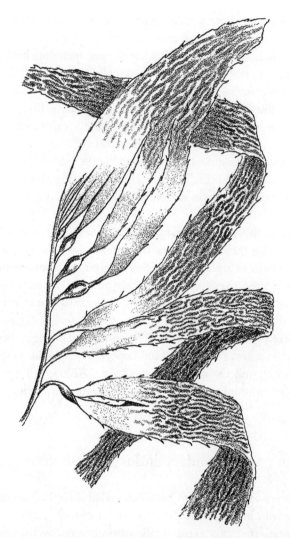

Figure 2.7 The brown alga *Macrocystis pyrifera* exemplifies the complex, multicellular seaweeds of the Brown algae. Drawing from Oltmanns 1905.

they ingested the chloroplasts from other algae. In addition, there are some dinoflagellates that actually have four membranes around their chloroplasts, suggesting that they stole chloroplasts from an organism that had previously stolen chloroplasts from still another organism ("tertiary endosymbiosis").

The brown chloroplasts of brown algae, diatoms and dinoflagellates are thought to have come originally from a red alga. Little is known about when or why the pigment shift took place, but it appears to have happened just once, as all brown chloroplasts are similar structurally and chemically. Different kinds of predatory cells have taken them in. Dinoflagellates, for example, share cellular characteristics

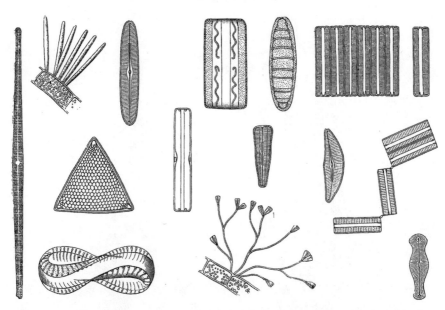

Figure 2.8 Diatom cells are enclosed in fitted glass shells resembling shoe boxes, and come in an astounding array of intricately ornamented shapes. Some stick together in colonies or filaments. Modified from Kerner & Oliver 1895.

with the group of predatory protists known as ciliates, which include the familiar *Paramecium*. Their chloroplasts are mostly of the brown type, but some are green, which they acquired secondarily by stealing them from green algae. Such events are still occurring (Gast et al. 2007).

In sum, though there are just three basic types of chloroplasts, many of the eukaryotic lineages, exhibiting varied cell structures, and including even some animals, have acquired one or more of them through primary endosymbiosis, secondary endosymbiosis, or through a mutualistic relationship (Fig. 2.10).

The peculiar history of red and green algae

Because of their likely common ancestry, red and green algae have sometimes been included with the land plants in the formal plant kingdom, a grouping more recently called the Archaeplastida ("ancient [chloro] plastids"). Land plants certainly arose from green algae, but the relationship between red and green algae, is still controversial.

The first verified algal fossils, estimated at 1.2 billion years of age, were of a red alga, *Bangiomorpha pubescens* (Butterfield 2000) (Fig. 2.11). Identification of these fossils is based more on cell wall structure than any remnants of pigments in them, and they strongly resemble some modern forms of red algae. The green

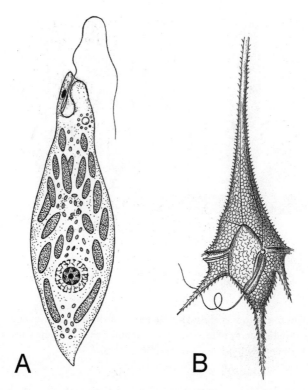

Figure 2.9 A. *Euglena* is an animallike protist without a cell wall, whose ancestors acquired chloroplasts through secondary endosymbiosis with a green alga; B. *Ceratium* is one of the diverse group known as dinoflagellates, which have acquired chloroplasts through both secondary and tertiary endosymbiosis involving mostly red and brown algae. It is the occasional population explosions of a dinoflagellate that cause red tide, which results in large-scale death of fish and other larger animals, due to a toxic byproduct of dinoflagellate metabolism. Drawings from Oltmanns 1905.

algae appear to be somewhat younger, first appearing as fossils about 1 billion years ago, and brown forms of algae are probably only half as old as them. There are various estimates that both red algae and green algae are older than the fossil record indicates, and that the first eukaryotes may have appeared as early as 2.7 billion years ago (Knoll 2014).

The chloroplasts of modern red algae contain accessory photosynthetic pigments called phycobilins, which include the reddish phycoerythrin mentioned above. Phycobilins are abundant in cyanobacteria, but not present at all in green and brown chloroplasts. This suggests that the first chloroplasts were like those of red algae, and that the novel pigments of green and brown algae came later.

There have been several prominent theories for how the green chloroplasts, and the green algae that possess them, came about. Among modern free-living photosynthetic bacteria, there are some, called prochlorophytes (or Chloroxybacteria, in the terminology of Margulis), that have the same general mix of accessory

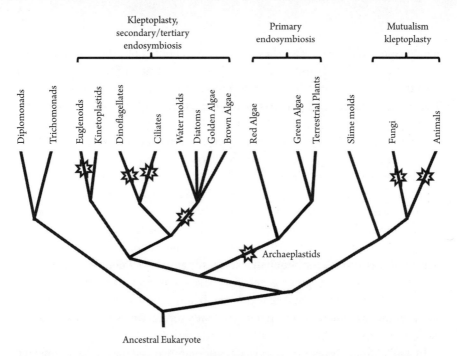

Figure 2.10 Within the broad tree of eukaryotic life, based on host cell characteristics, acquisition of photosynthesis through endosymbiosis or kleptoplasty has occurred in nearly every major clade, including the animals.

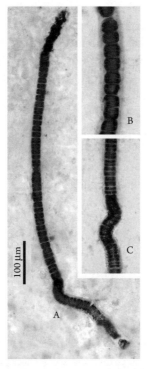

Figure 2.11 Fossils of *Bangiomorpha pubescens* are the oldest verified evidence of eukaryotic algae. They strongly resemble modern red algae of the genus *Bangia*. Photograph courtesy N. J. Butterfield.

pigments as the green algae: they lack the phycobilins of other cyanobacteria and have chlorophyll b and carotenoids like green algae. A few years back, this caused a lot of excitement. The prochlorophytes were thought to be quite distinct from cyanobacteria, and potentially the ancestors of green algal chloroplasts. It was proposed that the red algae originated through endosymbiosis with a phycobilin-bearing cyanobacterium, and the green algae through a completely separate endosymbiosis with a prochlorophyte.

Alas, the story is not that simple. With conflicting evidence from modern DNA analysis, the relationship between red and green algae has re-emerged as a controversial issue (see Graham et al. 2000 for a good summary). For example, a DNA-based analysis (Chen et al. 2005) showed that the several kinds of prochlorophytes may not be closely related to each other, nor to the chloroplasts of green algae. It seems also that chlorophyll b can easily appear as a mutant form of chlorophyll a. Alternatively, "green genes" may have evolved once and moved to other organisms through horizontal gene transfer. In any case, it seems likely that green photosynthetic pigments evolved independently several times as an adaptation to the wavelengths of light found in shallow surface waters.

In 2007, Adrian Reyes-Prieto and colleagues summarized the available comparative DNA studies for both chloroplast and nuclear genomes, and concluded that red, green, and brown chloroplasts were more closely related to one another than to any known cyanobacterium, and therefore probably had a common ancestor. They also concluded that there was most likely a single ancestral host cell that captured the cyanobacterium that became the grandfather of all chloroplasts. Those red chloroplasts evolved new color forms as their host cells adapted to different light environments.

Soon after, however, reports by Kim and Graham (2008) revealed that the relationships of red, green, and other algae are more complex than earlier thought. The red and green algae may have descended from a single host cell undergoing endosymbiosis with a single cyanobacterium or with several related cyanobacteria, or different host cells may have undergone endosymbiosis with different cyanobacteria. Nozaki et al. (2009) also concluded that red and green algae had a common origin but that the ancestors of brown algae were more closely related to green algae, but had lost their original chloroplasts before acquiring new ones through kleptoplasty. In other words, the issues of relationship among algae are far from resolved.

How did red algae lose their flagella?

Assuming that red and green algae did descend from a common ancestor, there arises another perplexing mystery about red algae, which lack flagella entirely. Because of their unique structure, it appears that flagella evolved just once early during the evolution of eukaryotes, and were then passed on to all subsequent eukaryotes. Flagella are virtually the same structurally in the sperm and other motile cells of animals, aquatic fungi, and algae.

Not one out of the thousands of species of red algae, however, has any trace of a flagellum in any part of its life cycle. Their cells even lack the basal structures (centrioles) in the cytoskeleton that are typically associated with flagella. Either they evolved from ancient eukaryotes before the invention of flagella, or for some reason flagella were lost or suppressed among early red algae. Most likely the latter is the case.

In the various phylogenetic analyses that have been done (e.g., Kim and Graham 2008, Nozaki et al. 2009), animals, fungi, and a variety of protists with flagella appear to have branched off of the eukaryote tree well before the red or any other kinds of algae evolved. So the common ancestor of red and green algae most likely had flagella like modern green algae, but chloroplasts similar to those of red algae. If so, the green algae retained flagella while evolving new chloroplast pigments, while the red algae retained more ancient chloroplast pigments while losing flagella.

The suppression of flagella must have happened early and very thoroughly in the ancestors of modern red algae. Textbooks tend to imply that the loss of flagella in red algae was accidental. It would be more satisfying, however, if we could explain the lack of flagella as an adaptation to some particular environmental challenge or opportunity, because despite their seemingly inferior sperm, the red algae are today both diverse and numerous.

Almost all algae other than the reds depend on flagellate cells to complete one or more phases of their life cycles, so it is hard to see what advantage would be gained by giving them up, and why not one modern red alga retained them. Why make it harder for sperm cells to move toward the eggs?

The sperm cells (spermatia) of red algae are quite tiny, and drift passively in the water until they contact an egg-bearing structure by chance. Eggs are retained on the female plant until fertilized, and the highly sophisticated egg-bearing structure has a sticky, fingerlike projection on which drifting sperm cells are caught. Moreover, once caught, sperm cells are pulled inside the apparatus to unite with the egg nucleus. The likelihood of that encounter between sperm and egg would however seem to be astronomically small. How did such a mode of reproduction lead to a large successful group of organisms?

The sperm cells of red algae are indeed tiny but produced in huge numbers. Since they lack flagella, they can forgo the machinery and energy reserves (typically provided by active photosynthesis) needed to operate flagella. Such inert specks are much cheaper to make than larger, active, flagellate cells, and their greater numbers offset the lower probability of success for each.

This is not an unusual strategy at all. Fungi rely on sheer numbers of tiny spores for their highly successful dispersal, and among seed plants, we can point to those that are wind pollinated rather than animal pollinated. Plants like grasses, oaks, and pine trees, produce huge quantities of pollen that is dispersed passively by the wind. Again the probability that any one pollen grain will be successful is small (compared with pollen carried directly from one flower to another by animals), but the sheer numbers insure success.

The trade-off between motility and sheer numbers works in environments where there are predictable currents or winds that can channel tiny reproductive bodies through a population. It is possible then that the ancestral red algae lived where currents favored the passive dispersal of tiny sperm cells, while the ancestors of green algae lived in quieter shallow water where sperm cells could get around only through their own effort. Both are positive adaptations to specific environments. The red algae did not lose their flagella—they got rid of them because there was a better way to move their reproductive cells around in a particular type of habitat.

Green algae and sexual reproduction

Multicellular terrestrial plants evolved from green algae and inherited sexual reproduction from them. Adapting the algal mode of reproduction for the dry land environment was a huge challenge, one that drove many of the evolutionary trends we will see in future chapters.

Sexual reproduction in algae and plants is fundamentally similar to that in animals. Two cells (gametes), most often in the form of sperm and egg, fuse, combining their genetic information into a single cell (a zygote). With two sets of chromosomes, the zygote is diploid. Sometime later, a special division (meiosis) occurs, which recreates cells each with a single (haploid) set of genetic information, but a set that is now a mixture of information from the two parents.

Why and how did this complex process of sexual reproduction evolve? Why not just multiply through ordinary cell division or asexual budding? Theories abound on this topic, which is still quite controversial (see Zimmer 2009, for an excellent review), but the benefits of the genetic mixing that results from sex are fairly clear.

There are organisms that reproduce only asexually, including some plants, but this practice results in populations of identical clones, not genetically new individuals. If a disease comes along, it is likely to wipe out the entire population, as would unusually cold weather or the arrival of a new predator. In a genetically variable population, however, chances are that some individuals would be more resistant to the disease, more cold tolerant, or more able to outrun the new predator. They would survive and pass their genes onto future generations. So the value of sex—the fusion of gametes followed by meiosis—is to provide a constant stirring of the genetic pot and populations better able to respond to environmental challenges.

The organized mixing of genetic traits from two individuals became possible only in eukaryotes after their cytoskeletons became adept at moving things around inside the cell. The genes of bacteria are for the most part lined up on a single circular chromosome. When a bacterium divides, its chromosome is duplicated and identical copies go with each new cell. Bacteria do have several ways of combining DNA from different individuals, but not through the fusion of sperm and egg. As you saw in the previous chapter, the process of horizontal gene transfer had an immense impact on the evolution of bacteria and of important processes like photosynthesis.

In eukaryotes, the genome is broken up into a set of distinct chromosomes, each with a specific part of the organism's total genome, and each of which can move independently. The process of mitosis evolved to insure that each cell got a copy of each chromosome during cell division. After the chromosomes duplicate themselves, they line up in the middle of the cell, and the two copies of each are pulled in opposite directions Fig. 2.12). It produces genetically identical cells and is the way unicellular organisms build up populations and multicellular organisms grow and develop.

Meiosis is similar to mitosis but contains a phase in which the chromosomes representing the same part of the genome (homologous chromosomes) from each parent pair up and then separate randomly to the opposite sides of the cell. This results in two single (haploid) sets containing different mixtures of chromosomes from the two parents. A second phase separates duplicates of each chromosome, resulting in four sets (Fig. 2.13). Further mixing can occur if chromosomes exchange pieces during the pairing phase, so potentially, each of the four nuclei resulting from meiosis can be genetically unique.

Meiosis can occur at different times after the union of sperm and egg, resulting in radically different life cycles. In most algae, meiosis occurs in the zygote itself, and the first haploid cells out of the zygote are usually equipped with flagella and chloroplasts and serve as zoospores. Zoospores can swim about for a bit, and if they end up in a suitable environment can multiply into a new population or multicellular individual. When gametes form again later, the process repeats itself. This series of alternating haploid and diploid cells is called a life cycle (Fig. 2.14).

If meiosis does not occur directly in the zygote, a diploid individual may develop. This is what happens in animals, where meiosis does not occur until adulthood when gametes are produced. Among algae and plants, there are even more possibilities, resulting in a complex array of life cycles. Life-cycle variations relate to different strategies for maintaining optimal genetic diversity among offspring, as well as some other advantages of going about life as diploid individuals.

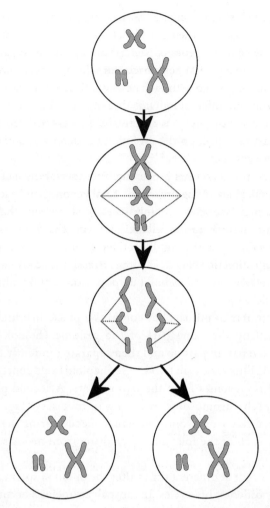

Figure 2.12 Mitosis is the process by which identical copies of each chromosome are distributed to two new cells during ordinary cell division.

Gametes and zoospores

For genetic mixing, it is not sufficient for a sperm cell to combine with the nearest available egg. That egg could be from a sibling, or even from the same plant. Such inbreeding, generation after generation, would be little better than cloning. Finding a more exotic mate will serve its descendants much better. The reproductive process therefore must accomplish two vital functions: 1) combining sperm and egg; and 2) ensuring that the sperm and egg are genetically different. The first is an intimate, close-up event, but the second requires some travel, and except for the spermatia of red algae, sperm cells really can't go very far.

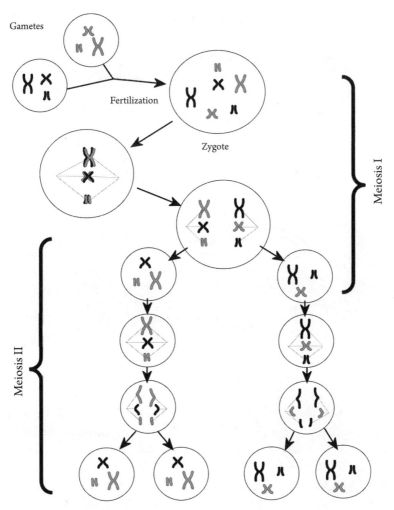

Figure 2.13 Meiosis occurs in specialized diploid cells as part of the sexual reproductive cycle. The process mixes the sets of chromosomes received from the parents, resulting in a new generation of genetically variable haploid cells—gametes in animals, spores in plants.

Animals move about to select mates, and thus insure that their gametes mix with gametes from genetically different individuals, but in algae, the traveling part requires a special go-between, the zoospore. The job of a zoospore is not to fuse with another zoospore, but rather to disperse away from its parent, to mix with another colony or to form a whole new colony somewhere. This special event is the dispersal phase. The zoospores do the job of finding suitable mates for individuals before they are born. Remember that zoospores come from the zygote, which came from the fusion of gametes. So the sexual part occurs before the production of new zoospores, between genetically different neighbors brought in by the previous generation of zoospores.

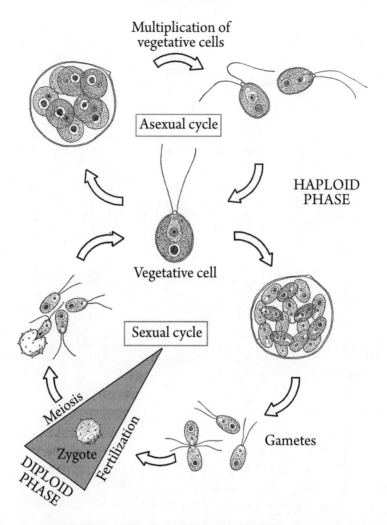

Figure 2.14 *Chlamydomonas* exhibits a simple life cycle in which populations consist of individual haploid cells that multiply asexually by simple cell division (mitosis). The diploid phase comes about as specialized cells from genetically different strains serve as gametes and fuse together. The zygote undergoes meiosis after a period of dormancy, generating haploid zoospores that become vegetative cells. Redrawn from Haupt 1953.

Contrast this again with the more familiar life cycle of animals, in which the zygote develops into a multicellular diploid individual. The zoospore phase is bypassed, and meiosis is delayed until the animal matures, and then it produces gametes. The multicellular adult finds its own mate and delivers the sperm to the egg.

Between the life cycles of simple algae and those of animals, we find a peculiar middle ground, in which there is both a haploid body and a diploid body. An example is the marine green alga, *Ulva*. The diploid zygote does not undergo meiosis but instead develops into a flat, multicellular body identical to that of the haploid phase. The haploid individual produces gametes, like other green algae, and the diploid individual produces zoospores through meiosis, also as expected, but only after forming a multicellular structure (Fig. 2.15). A life cycle with both haploid and diploid multicellular phases is said to go through alternation of generations. This is uncommon in green algae, but we will see in the next chapter that it is the norm in higher plants.

Just exactly why *Ulva* undergoes alternation of generations is a matter of speculation. The diploid form is able to produce a large number of zoospores, while the haploid form can produce a large number of gametes. Perhaps some environmental conditions favor gamete production (local conditions are right for gametes finding each other over short distances), while others favor zoospore production (conditions favor long-distance dispersal of zoospores).

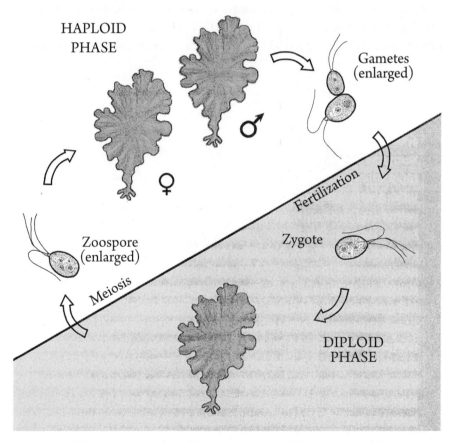

Figure 2.15 The marine green alga *Ulva* exhibits an equal alternation of generations; the diploid sporophyte and the haploid gametophyte are identical in size and appearance. The sporophyte produces motile zoospores through meiosis, while the gametophyte produces motile gametes through mitosis. Redrawn from several classical sources.

Populations of *Ulva* may contain a mix of haploid and diploid individuals and be ready to do either.

So there are three general life cycles among the major groups of life: those in which ordinary cells are all haploid (most green algae), all diploid (animals), or alternate between haploid and diploid phases (some algae, higher plants) (Fig. 2.16). Fusion of gametes and meiosis are the two events that transition from one phase to the other.

Note that in many algae, *Ulothrix*, *Chlamydomonas*, and *Ulva*, for example, gametes all look the same (isogametes). There is no sperm or egg, but rather gametes from different strains that differ only chemically. Gametes look like slightly smaller versions of ordinary cells, but are programmed to fuse with another gamete rather than to go about the usual business of photosynthesis and multiplication. This may be the way all gametes were at first.

With further refinement, gametes became specialized as small, motile sperm and large, nonmotile eggs. Such gametes are found in *Volvox* (Fig. 2.17), a colonial cousin of *Chlamydomonas*, and many other green algae, including the ancestors of land plants. Sperm cells in plants typically have flagella that pull the cell

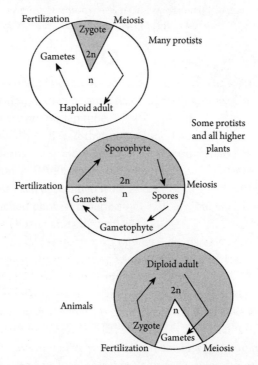

Figure 2.16 All sexual life cycles include an alternation of haploid and diploid phases. In some algae and all plants, there are both haploid (gametophyte) and diploid (sporophyte) multicellular phases or "generations." Meiosis produces spores, and gametes are produced through mitosis.

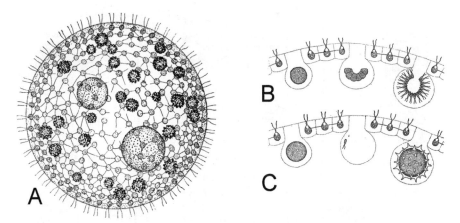

Figure 2.17 Volvox (A) forms a spherical colony with cells connected by thin rods; certain cells on the surface of the spherical colony divide to form packets of sperm cells (B), while others (C) develop into a single large egg. The egg remains in its chamber until fertilization, after which it moves to the inside of the sphere and develops into a new colony. Once several new colonies have formed within, the original colony breaks apart to release them. Drawings from Coulter 1910 (A) and Smith 1938 (B).

forward, while the animal sperm has a single rear-mounted flagellum that pushes it forward, indicating that sperm and egg evolved independently in plants and animals.

In a few green algae, such as *Spirogyra* and its relatives, motile gametes are lacking. Instead, conjugation tubes form between mating filaments and nuclei travel from one to the other to achieve sexual fusion (Fig. 2.18).

As we'll see in the next chapter, reproduction by active sperm cells continued as plants moved onto the land, but spores had to be transformed for travel through the air. This added some new complications to the life cycle and new vegetative forms.

The green algae and land plants

By the end of the Precambrian Era, roughly 600 million years ago, the oceans were populated by a great variety of eukaryotic algae, representing everything from simple planktonic forms to giant kelp that formed underwater forests. These algae contributed to the final buildup of oxygen (and ozone) in the atmosphere to roughly current levels, preparing the way for the movement of plants and animals onto the land. The green algae already predominated in shallow freshwater habitats, making use of the full spectrum of sunlight available, and so it was natural for them to move into soil and other substrates moistened by fresh rainwater.

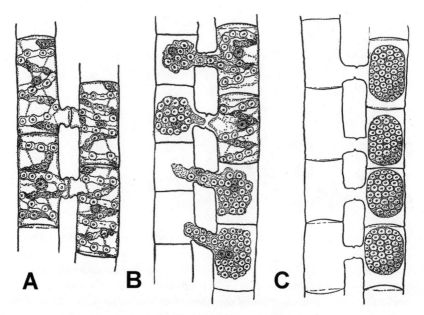

Figure 2.18 In *Spirogyra* and some other algae, fertilization is by means of conjugation, rather than sperm and egg; during conjugation, a tube forms between adjacent filaments, and the contents of the cells in one filament pass into the cells of the other filament, where they merge into diploid zygotes. Drawing from Haupt 1953.

Green algae have exploited every conceivable growth form, including unicellular, filamentous, and colonial forms, as well as three-dimensional seaweeds. They show great variation in size, in cell and chloroplast structure, and in their sexual life cycles. Many green algae have independently evolved into three-dimensional seaweeds. Which of them first came ashore?

The first green algae were unicellular and simple colonial forms (Fig. 2.19), and similar species still make up an important part of the floating plankton community in both freshwater and saltwater environments. They share this role with diatoms, dinoflagellates, euglenoids, and others. They often form remarkably symmetrical colonies, as in *Volvox* (See Fig. 2.17).

An important step toward plantlike growth forms was for cells to stick together after division, sharing the common wall between them, and forming long, one-dimensional filaments. Common examples include *Spirogyra* (see Fig. 2.18) and *Ulothrix* (Fig. 2.20). More complex growth forms arise when cells divide in more than one plane, sticking together in two-dimensional sheets, as in *Ulva*, (see Fig. 2.15), or become increasingly plantlike, with stems and clusters of photosynthetic branches, as in *Draparnaldia* (Fig. 2.21). More bizarre forms, such as *Caulerpa* (Fig. 2.22A), the organs of which consist of giant, branched, multinucleate cells, the umbrellalike *Acetabularia* (Fig. 2.22B), or the netlike *Hydrodictyon* (Fig. 2.23), are evolutionary side shoots unlike higher plants.

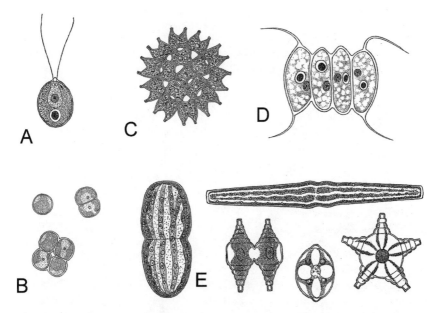

Figure 2.19 Many green algae are free-floating individual cells or small colonies, including motile forms like *Chlamydomonas* (A), and nonmotile forms such as *Protococcus* (B), *Pediastrum* (C), *Scendesmus* (D), and various desmids, (E). Drawings from Haupt 1953 (A, B), Brown 1935 (C), and Oltmanns 1905 (D and E).

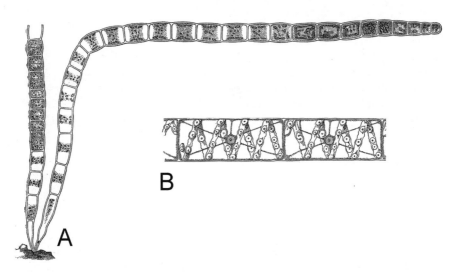

Figure 2.20 A. *Ulothrix* is a filamentous green alga that attaches to rock, wood, or other substrates via a basal holdfast; B. *Spirogyra*, with its unusual spirally wound chloroplasts, often forms into a free-floating mat. Drawings from Kerner & Oliver 1895 (A) and Haupt 1953 (B).

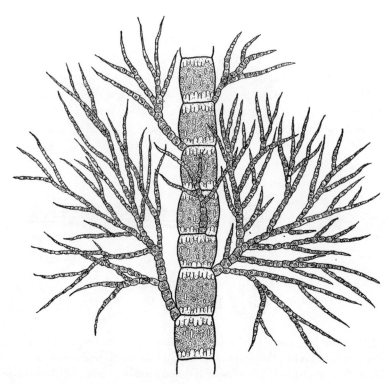

Figure 2.21 *Draparnaldia* is a branching filamentous green alga with a main stem of rather large cells and bushy lateral appendages of much smaller cells. Drawing from Haupt 1953.

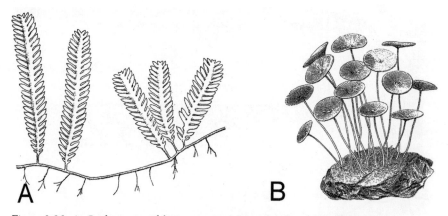

Figure 2.22 A. *Caulerpa crassifolia* is a marine green alga that forms feathery stalks attached to the sea bottom; the entire colony is one multinucleate cell (coenocyte). A cold-hardy form of this usually tropical seaweed has recently invaded the Mediterranean Sea, crowding out native vegetation and upsetting the food chain (*Caulerpa* is toxic to most fish); B. *Acetabularia* is another marine green alga, unique for its satellite-dish-shaped light-gathering antennas atop long stalks. Drawings modified from Smith 1938 (A) and Oltmanns 1905 (B).

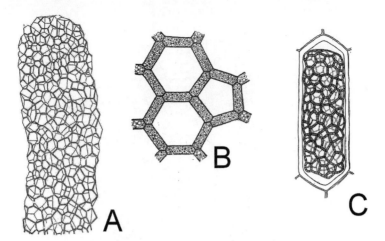

Figure 2.23 A mature colony of *Hydrodictyon* (A) consists a saclike network of cells connected in polygons; a magnified section of the colony (B) showing the interconnected cells; miniature new colonies (C), which later expand to full size, may form asexually within cells of the parent colony or through sexual reproduction. Drawings from Kerner & Oliver (A, B) and Haupt 1953 (C).

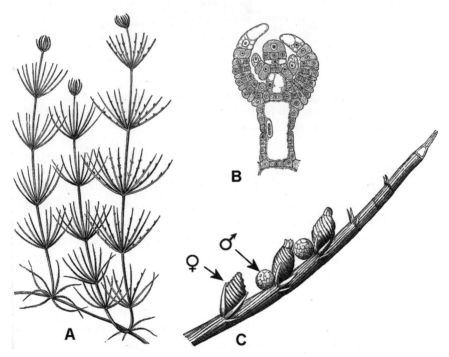

Figure 2.24 The architecture of *Chara* (A) is remarkably like that of higher plants, with distinct nodes, internodes and leaflike appendages; new appendages and internodal segments are produced from embryonic cells at the tip (B); internodes consist of a single enlarged cell surrounded by a jacket of smaller cells. Male and female gamete-producing chambers (C) are borne on special branches. Source: Kerner & Oliver 1895 (A, C) and Haupt 1953 (B).

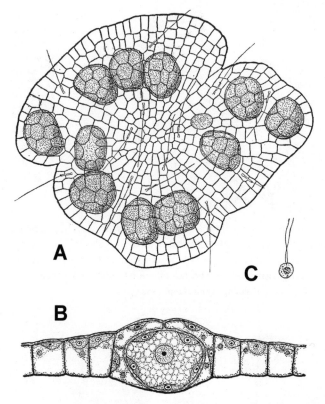

Figure 2.25 *Coleochaete* species form a flat multicellular thallus, with large eggs produced within specialized cells (A,B); the eggs are fertilized by flagellate sperm cells (C) produced on a different plant. Drawings from Haupt 1953.

In algae where cells stick together as either long filaments or three-dimensional seaweeds, we see the emergence of an important plant characteristic: indeterminate growth. These organisms can continue to grow over time, adding new tissues and organs, usually in localized growth areas. This can result in extensive interconnected light-gathering complexes of indefinite size, shape, and age. Such complexes, whether a tangled mat of filaments or an organized seaweed, typically anchor themselves to one spot, a characteristic of most higher plants.

In the group known as the Charophytes, we see true multicellular structure and differentiation of specialized stems and photosynthetic organs. *Chara*, for example, has slender, photosynthetic branches radiating out from distinct nodes along thick stems (Fig. 2.24). It would seem to be a starting point for mossy plants to evolve on land. Neither *Chara* nor any of the other three-dimensional green algae mentioned above was the actual ancestors of land plants, however, not even their closest relative.

That honor, according to modern phylogenetic studies, belongs to *Coleochaete*, a cousin of *Chara*. The typical form of *Coleochaete* species is a flat, roundish, two-dimensional sheet (Fig 2.25) that grows outward from the center, quite similar to the simplest terrestrial plants: hornworts and liverworts. We take up their story in the next chapter.

Figure 3.1 Damp habitats such as this temperate rain forest in the Hoh River Valley, Washington, host a multitude of mosses and other bryophytes that cling to every available surface.

3

Plants Invade the Land

The Hoh Rainforest sits in a narrow valley on the west side of Olympic National Park in Washington State. With over 12 feet of rainfall per year, it is one of the wettest "dry land" environments on the planet. Thriving virtually everywhere in this forest—on tree trunks and branches, on the ground, on rocks, in the streams—are spongy masses of the humble plants known as mosses (Fig. 3.1). Other than a vague awareness of this green carpet, most people pay little attention to mosses, or to their even less conspicuous relatives, the liverworts and hornworts. But in the Hoh Rainforest, they are hard to ignore as they blanket nearly every available surface, even hanging as festoons or curtains from overhead branches.

Instead of competing with the trees around them, mosses, liverworts, and hornworts make a living from the leftover light that trickles down through the forest canopy. As such, however, they are extremely successful. Though they may all look the same at first, under a hand lens or microscope, these tiny plants reveal an astonishing variety of structural details, reflecting some 500 million years of evolution. With over 23,000 species, they are more diverse than any other group of land plants except the flowering plants.

Though they are usually inconspicuous, mosses make up the dominant vegetation in certain parts of the world. Vast areas of northern Canada and Russia are covered by wet, acidic bogs filled with mosses of the genus *Sphagnum*. These mosses seem to be the only plants that can tolerate and thrive under these cold, wet, highly acidic conditions. Living in such an environment, sphagnum moss has natural antifungal properties and is highly resistant to decay. Its dead remains may accumulate for many centuries. The resulting deep beds of peat are mined by indigenous people for fuel, are sometimes even burned in power plants, and are exported as a horticultural potting medium. Fresh sphagnum was used in the past as a sterile dressing for wounds, and it is still used in seed-starting mixes to prevent fungal attack on seedlings. The peat preserves other organisms that have fallen into the bogs, including the occasional unlucky Stone Age hunter, dug up thousands of years later, with skin and clothing intact.

The embryophytes

Mosses, liverworts, and hornworts are three distinct lineages descended from the most ancient of land plants, and together are referred to informally as the nonvascular plants. The more massive plants of the forest—the trees, shrubs, and larger herbs—are the vascular plants, which constitute a fourth lineage of land plants (Tracheophyta). Together, these four lineages constitute the Embryophyta—a name generally synonymous with "true plants" or the plant kingdom. The embryophytes get their name from the fact that eggs are fertilized and develop into young embryos in special chambers, where they are nourished by the parent and protected from drying out.

The nonvascular plants lack the sophisticated, organized internal plumbing, or vascular tissues, found in their larger cousins. Referring to them by that name, however, is like calling reptiles "non-birds," and gives no clue as to the nature or great evolutionary success of these plants. They are alternatively referred to as bryophytes ("moss-plants"), though one must take care not to confuse that term with "Bryophyta," which in formal classification refers to just the mosses, and excludes liverworts and hornworts. What the three lineages of bryophytes have in common, aside from the lack of vascular tissues, is small size, a generally high tolerance of desiccation, and a sexual life cycle that is more like that of green algae than that of higher plants.

Mosses (Bryophyta), which have miniature stems and leaves, and often grow upright like higher plants (Fig. 3.2), range in size from a barely discernible green fuzz to cushiony masses several centimeters thick (or in a few species, as high as 60 centimeters). Liverworts (Marchantiophyta) (Fig. 3.3A and B) and hornworts (Anthocerophyta) (Fig. 3.3C) are generally even less conspicuous, lying prostrate on the ground, sometimes underneath a canopy of relatively larger mosses, or forming thin films on leaves or bark. The more "massive" liverworts, like *Marchantia*, which are staples in botany labs, have bodies one centimeter or more wide. *Marchantia* and *Anthoceros* take the form of flat, forking ribbons or sheets; forms of an indefinite body type referred to as a thallus. Other liverworts are "leafy": tiny creeping plants with flat, leaflike outgrowths on slender stems (Fig 3.3B).

With dividing cells concentrated at the front ends of the ribbonlike thalli, early plants resembling modern liverworts extended themselves horizontally across the soil surface, occasionally forking to form branch ribbons. In this way they could spread outwards into fresh terrain and expand into extensive colonies. Such growth and expansion can continue indefinitely—something we call indeterminate growth. Older parts of the colony eventually disintegrate, resulting in fragmented and isolated patches that form new centers of clonal expansion. Mosses have stems with leaves produced at their active tips. The dividing cells at those stem tips constitute an early version of the apical meristem, the localized growth center characteristic of all higher plants.

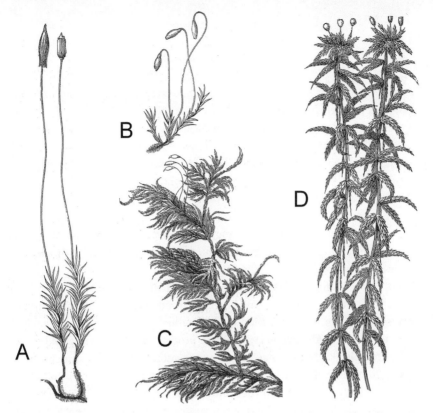

Figure 3.2 Mosses are quite varied in their form. Leafy stems are typically packed close together, and rhizoids provide anchorage and absorption. A. *Polytrichum commune*; B. *Bryum caespiticium*; C. *Hylocomium splendens*; D. *Sphagnum palustre*. Drawings from Kerner & Oliver 1895.

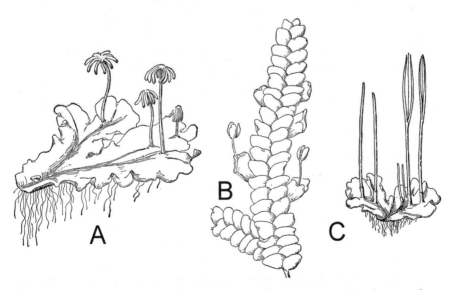

Figure 3.3 Liverworts have varied forms but generally remain pressed to the ground or other substrate. *Marchantia* (A) consists of a flat, branching thallus; *Porella* (B) has branching stems covered with tiny leaflike appendages; *Anthoceros* (C) is a hornwort, which also grows as a flat thallus. Drawings from Coulter et al. 1910.

The first colonizers

Life evolved in the seas and remained in the water for nearly 3 billion years. Organisms that lived in fresh water prior to about 500 million years ago were tantalizingly close to the vast untapped real estate of the terrestrial world, where light, minerals, and carbon dioxide were more abundant, and the competition for these resources virtually nonexistent. Aquatic organisms are bathed constantly in water. They need no specialized structures for absorbing it, and have no need to store or conserve it. But in the terrestrial environment, obtaining and maintaining appropriate body moisture is a real challenge. Multicellular plants, animals, and fungi faced these challenges and ultimately triumphed; for today, the terrestrial environment is teeming with life, even where it is quite dry.

Who were the first terrestrial organisms? Most likely, desiccation-tolerant bacteria and other microorganisms inhabited mud and other intermittently wet places on land prior to the invasion of multicellular organisms, but probably for not too long before. The ozone layer, which protects modern terrestrial life from lethal ultraviolet radiation, may have been insufficient for that function until around 500–600 million years ago. Low atmospheric oxygen also inhibited the evolution of larger animals and their movement onto land until about the same time. Perhaps the first multicellular organisms on land were fungi that fed on dead aquatic animals and algae that washed ashore. Lichens, the symbiotic association of fungi and algae or cyanobacteria, were also among the earliest land organisms. Though the oldest fully verified lichen fossils are only about 400 million years old, there is some evidence of their existence before 600 million years ago (Yuan et al. 2005).

Some green algae also invaded the terrestrial environment once the threat from ultraviolet radiation subsided. Though there is likely never to be any evidence of their existence at that time, we know from organisms alive today that certain algae can thrive in relatively dry habitats. There are actually a surprising number of green algae that are permanent residents of the terrestrial environment. Some live in areas that remain moist, like soil, rocks near waterfalls, or the leaves and bark of tropical trees, but there are others that occur in places that are often bone-dry.

Air is a terrible medium for living cells. Just as clothes hung out on a line dry within a few hours, dry air can suck the water, and the life, out of an unprotected cell. In Florida, where it may be quite dry for days or even weeks, there are species of unicellular green algae that can be found on walls, tree trunks, and screened pool enclosures. But like their aquatic relatives, these tiny cells have no means of storing or conserving water. Instead, they have evolved the ability to become completely desiccated (dehydrated) without suffering any damage.

Cells of several species of *Chlorella* and *Chlorococcum,* for example, can lose 97% of their water and then resume normal growth when wetted again (Chen and Lai 1996). They occasionally undergo sexual reproduction, which you can imagine is difficult with only fleeting episodes of hydration to work with, but are able to multiply asexually, forming thick coatings on suitable surfaces. Dried cells ("spores") disperse through the air to other suitable sites, often over great distances, incidentally, contributing to the allergenic pollution of the air. Their survival strategy is clearly to tolerate desiccation rather than to resist it.

The first multicellular plants

All four lineages of embryophytes appear early in the fossil record, though the vascular plants most likely did come a little later. Their precise phylogenetic relationship is still uncertain. The hornworts (Renzaglia et al. 2000) and liverworts (see Ligrone et al. 2012) have each in turn been identified as the most ancient group, with mosses and vascular plants emerging from them. All are part of the story of the invasion of dry land, and their different strategies for survival and reproduction will be explored in this chapter. Much more will be said about the vascular plants in the chapters to come.

Some meager fossil evidence, only spores in fact, suggests that the first multicellular plants came onto the dry land in the Ordovician period, some 475–500 million years ago. The spores we find in these ancient rocks are similar to those of modern liverworts, and the plants that produced them were probably little more than creeping ribbons of green tissue confined to wet soil and other damp surfaces.

There is no doubt that the land plants evolved from green algae, specifically from a group called the charophytes, for they share many properties with them: the same set of photosynthetic pigments, similar cell walls and mode of cell division, similar starchy food reserves, and similarities in their sperm and eggs. DNA comparisons confirm the relationship. The charophyte genus *Coleochaete* is probably the closest living relative of land plants. Some species of this genus grow as a flattened disk similar to the thallus of a liverwort or hornwort, as well as to the corresponding gamete-bearing thallus of ferns (Fig. 3.4). *Coleochaete* will be referred to frequently as a model of what land plant predecessors were probably like.

In coming onto dry land, these descendants of green algae faced the same two problems as the first terrestrial animals: maintaining body moisture and getting sperm to egg. Land plants had a third problem, however: dispersing their progeny. Under the best of circumstances, a sperm can travel but a short distance to find an egg cell, and if the resulting offspring stayed close to home as well, it would lead to inbreeding and negate the genetic advantage of sexual

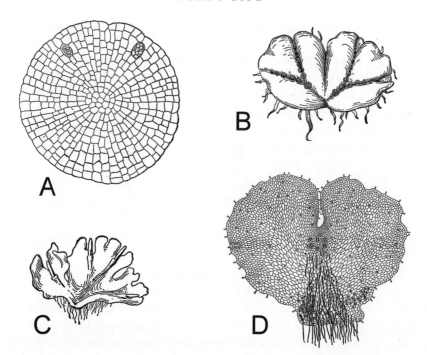

Figure 3.4 The green alga *Coleochaete* sometimes grows as a flat disk, expanding at the edge, resulting in a thalloid growth form similar to that of many liverworts (B) and hornworts (C), as well as to the gamete-producing bodies (gametophytes) of ferns (D). The dark objects in A are fertilized eggs, or zygotes, which will undergo meiosis to produce zoospores; in B, C, and D, the zygotes will develop into multicellular spore-producing structures (sporophytes). Drawings from Coulter et al. 1910.

reproduction. Early land animals could move about and mate with genetically different individuals, potentially even migrating to different populations. Early land plants had to find ways not only to survive and reproduce, but also disperse their progeny while sitting in one spot. The solution was to produce airborne spores, a process that greatly complicated the life cycle of land plants compared with their aquatic ancestors.

Evolutionary botanist G. Ledyard Stebbins (1974) emphasized that all plants face these three separate areas of adaptation: vegetative survival, sexual union (including pollination in higher plants), and dispersal (fruit and seed dispersal in higher plants). These three adaptive regions often proceed at different paces and may seem to lead in different directions. In early land plants, the first major changes concerned vegetative survival and then dispersal (spore production). The sexual process by which sperm unites with egg changed relatively little at first. Sperm still had to swim to the eggs through the water, a fact that did not change until the process was brought "indoors" in the first seed plants.

Tolerating desiccation

As remarkable as their success is, terrestrial algae live an intermittent existence, photosynthesizing and reproducing during wet periods that are sometimes extremely brief, and sitting in a state of suspended animation the rest of the time. Early multicellular plants faced the same challenges as they came onto the land. Some may have stayed close to water, but others became remarkably desiccation-tolerant, often living next to unicellular green algae and sharing their intermittent lifestyle.

For example, the variety of modern bryophytes living on tree trunks, dry sand, and rocks in central Florida, where there are prolonged dry periods, is quite amazing. In the spring, weeks may go by without a drop of rain, and temperatures can be up into the high 80's (F) every afternoon. But the plants do not die. They are shriveled up, bone dry, but still green. One can take up a tiny sprig of a dry moss or liverwort, add a drop of water, and it will spring back to life within seconds, the cells fully turgid, and the chloroplasts a vibrant green. Some mosses living in deserts can revive after 20 years of total desiccation.

The few plants able to survive, by taking advantage of brief moments of above-freezing temperatures, in arctic and Antarctic regions are also bryophytes. Recently, mosses estimated to be frozen for 400 years in northern Canada were recovered and brought back to life (La Farge et al. 2013).

Other bryophytes live in more hospitable environments and so can grow more luxuriantly. Some are fully aquatic, though they probably adapted to the water secondarily, just as the terrestrial ancestors of whales and dolphins did. Like aquatic mammals, the aquatic bryophytes carry vestiges of their terrestrial ancestry, primarily features of their reproduction that we will explore shortly.

The vast majority of bryophytes that live exposed to the air, even in rain forests, are subject to some degree of desiccation and are adapted to withstand it. Actual internal water storage, other than that held within ordinary cells, is uncommon in bryophytes. In most, leaves are only one cell thick and dry out quickly. The many species of *Sphagnum*, which constitute the greatest biomass of mosses on Earth, are exceptions. They possess specialized water storage cells intertwined with a network of photosynthetic tissues (Fig. 3.5).

Most mosses rely more on what might be called a "penguin strategy," not for retaining heat but for storing water. The penguins that spend winters in Antarctica stand upright, closely packed together to conserve heat. Mosses similarly pack together close enough to hold drops of water between them, or at least to maintain a stagnant matrix of humid air. They may form spongy mounds of upright stems or tangled mats lying against the ground. Stepping into a moss colony after a recent rain storm is literally stepping into a wet sponge. Liverworts and hornworts typically remain flat against the ground, maintaining moisture beneath them. Such

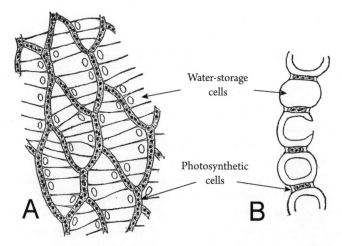

Figure 3.5 Sphagnum leaves are only one cell thick, but consist of two different kinds of cells: the narrow photosynthetic cells form a network around patches of large, clear, water-storage cells. Each storage cell has a large pore, through which water can enter. A, surface view; B, cross section. Drawings from Smith 1935.

multicellular, three-dimensional colonies of plants can remain active much longer than single-celled algae clinging to a rock or the side of a tree.

The external plumbing of bryophytes

Most nonvascular plants not only lack internal water storage, but also lack any system for internal transport. They have no roots and absorb water and minerals directly through their exposed surfaces, from either raindrops or the soil. This happens primarily on the outside of the plant in bryophytes. Water seeps from lower, wetter parts of the plant to upper, drier parts, much in the same way that a paper towel absorbs water. The physical process by which water defies gravity in this way is due in part to something called capillary action.

Capillary action results from the attraction of water molecules to one another (cohesion) as well as to physical surfaces (adhesion) such as the cellulose of plant cell walls. Because water molecules are asymmetric, the oxygen side of the molecule has a slight negative charge, and the hydrogen side a slight positive charge (Fig. 3.6), and so they act like tiny magnets. As water evaporates from the upper surfaces of a plant, water molecules are drawn upward to replace them. This then pulls on the chain of water molecules below. Thus the water molecules that are lost through evaporation are replaced by a continuously flowing film of water moving up from the damp soil or from the interior of the spongy mat. The

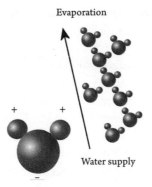

Figure 3.6 Because of unequal sharing of electrons between the two small hydrogen atoms and the larger oxygen atom, water molecules are polar, which means they have a positive end and a negative end. This causes them to stick together and to charged surfaces; and to be pulled upward as molecules evaporate from the top of the column.

upward movement of water in plants due to capillary action and evaporation is called transpiration.

Movement of water through a plant is thus a passive, physical process. No living, metabolic process is required. For example, the lower parts of sphagnum moss are dead, but their cellulose walls continue to serve as wicks for the upward movement of water. As long as there is sufficient moisture in the soil or within the colony, the upper parts of the plants will be watered. Water from this upward capillary movement, as well as directly from raindrops, is absorbed by the cells of the plant as needed. Plants swell as cells take up water through osmosis and develop turgor pressure.

Osmosis cannot be fully explained in a few words, but fundamentally, it is the result of water following concentration gradients. Water is less concentrated inside cells than out, because of a dense concentration of mineral ions and other particles (solutes) in the cell. Intermixed molecules of water and solutes tend to diffuse until uniformly mixed. Cell membranes allow water to enter and exit a cell freely, but prevent solute particles from leaving, so there is a net movement of water molecules inward, which would continue until concentrations were equal inside and outside the cell. But that never happens.

An unprotected animal cell placed in distilled water will swell until it bursts. In plant cells, the cell walls restrain the expansion, and so the constant influx of water raises the internal pressure. This is turgor pressure, which is normal in fully hydrated plant cells. At equilibrium, turgor pressure forces water out at the same rate it enters.

Turgor pressure creates the stiffening support that holds soft plant parts upright. As plant tissues lose their turgor pressure, they become limp and shriveled, as in a piece of lettuce sitting out on a table. The leaves of mosses sometimes twist, curl, or fold as this drying occurs, but they spring back into their active shape as water again becomes available.

Water movement in most nonvascular plants is primarily on the surface and proceeds slowly from one cell to another. Only a few of the larger mosses have some internal water transport, in the form of primitive water-conducting cells, called hydroids, in the center of their stems. They also have a greater degree of waterproofing cuticle on their leaves. In this way, they resemble the vascular plants, which have internal plumbing and a much greater mass of internal water-storage tissue. Why no moss proceeded further to become a larger plant will become clear at the end of this chapter.

Vascular plants maintain continuous internal supplies of water

Most of the plants we see around us today are vascular plants, with comprehensive systems of internal transport. They are typically larger and more complex than bryophytes. Some are trees over 100 meters tall. They include club mosses, ferns, gymnosperms, and flowering plants. Simple external water transport and desiccation tolerance are insufficient for these larger sizes. With an internal system of very narrow tubes, however, capillary action and evaporation can maintain an existing stream of water column in a tall tree. Evaporation from the leaves exerts a powerful pull on the collective chains of clinging water molecules extending all the way down to the roots. Within the narrow tubes, water molecules cling tenaciously to one another and continuously move upwards. However, such a massive internal stream cannot be restored if it should run dry.

In a dehydrated tree or even a much smaller shrub or herb, the connection between the evaporative force at the top and the water supply at the bottom is severed. A renewed water supply in the roots can rise only so high through capillary action before the drag of gravity brings it to a halt. So, except for some relatively small forms, vascular plants cannot recover from severe dehydration. The small epiphytic resurrection fern, *Pleopeltis polypodioides*, which is common in Florida (Fig. 3.7), is one of the few able to do so. Vascular plants thus shifted from a strategy of desiccation tolerance to one of desiccation prevention, and they are characterized by adaptations for improved absorption, internal transport, storage, and control of water loss.

Vascular plants protect themselves first of all with a hard, waxy cuticle, which is thicker and more fully impermeable than the modest cuticles found in bryophytes. The cuticle is secreted by the tightly packed outer layer of cells called the epidermis. Since vascular plants have roots, which do most of the

Figure 3.7 The resurrection fern, *Pleopeltis polypodioides*, grows on tree and palm trunks in Florida. These ferns can be completely dehydrated for weeks and revive after a rainstorm.

absorption of water, the leaves and other surfaces can be more thoroughly waterproofed than those of nonvascular plants. A thick cuticle can seal a plant off almost completely from the external environment, and that is pretty much what desert cacti do under dry conditions. Cacti store considerable quantities of water in their tissues, and they replenish that store through their roots whenever the soil is moistened by the rain. They can remain alive for months or even years between rare desert storms, but they cannot grow much during that time.

Growth requires an input of carbon dioxide in order to make carbohydrates, so all vascular plants, even cacti, must open themselves up sometimes

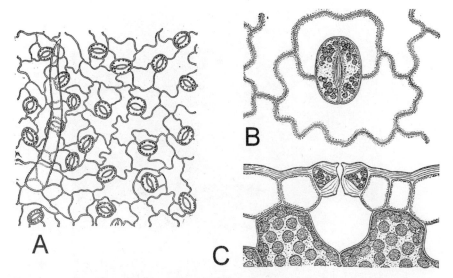

Figure 3.8 The epidermis of vascular plants and some bryophytes have closeable pores called stomata (A), which consist of two sausage-shaped cells containing chloroplasts (B). In cross section (C), one can see that the epidermal cells have a thickened outer wall, and that the stomata sit above chambers where gases can diffuse into the cells. Drawings from Brown 1935 (A) and Ganong 1916 (B).

via tiny pores called stomata (Fig. 3.8). This allows water vapor to diffuse out at the same time carbon dioxide is diffusing in, seeming to defeat the original purpose of the cuticle! But stomata can be closed when the threat of dehydration is highest or when absorption of carbon dioxide is not needed. Most plants close their stomata at night when photosynthesis is halted. Cacti, however, reverse that, opening to admit carbon dioxide at night, and closing during the heat of the day. Carbon stored at night is then released for use during the daytime. Cacti are highly adapted for desert existence, but the loss of water through stomata is a compromise faced to some degree by all vascular plants.

The unavoidable loss of water through transpiration is clearly a constant threat to plants, but that threat is balanced by some benefits that are not obvious at first. Evaporation of water from the leaves has a cooling effect, like that of evaporative coolers in greenhouses or the sweating of animals. Evaporative cooling in leaves exposed to bright sunlight prevents overheating and damage to the photosynthetic machinery. In addition, as transpiration pulls water upward in the plant, dissolved minerals are pulled up with it, and so this may be essential for keeping the foliage well nourished.

The transport of water was internalized in early vascular plants through the evolution of specialized, hollow conducting cells called tracheids ("vascular" and "trachea" both refer to vessels or tubes) (Fig. 3.9). A continuous strand or mass of tracheids is called xylem. Tracheids at maturity consist of thickened cell walls reinforced with lignin, a resinlike material that penetrates, seals, and strengthens the walls. The protoplast of the tracheid dies once the wall is formed, leaving a hollow tube. Tracheids are connected to one another by means of thin areas in the wall, called pits, that serve as filters. Water can flow freely through pits, but debris and air bubbles cannot. Early vascular plants contained strands of tracheids running through the center of each stem, and the trunk of a redwood tree is mostly a solid mass of these cells.

A special food-conducting tissue called the phloem evolved in parallel with the xylem. Dissolved sugar is actively pumped into the phloem in the regions where sugar is being produced (leaves) or released from storage. This causes water to move into the cells via osmosis, raising the turgor pressure in the phloem cells. In other parts of the plant (e.g., the roots), sugar is actively removed from the phloem, and water follows, resulting in reduced pressure. Sugar-laden phloem sap then flows from the region of high pressure to the region of low pressure. The flow of materials in the phloem is thus reversible, as roots may receive sugar during periods of photosynthetic activity and later release that sugar to fuel spring growth. The rise of maple

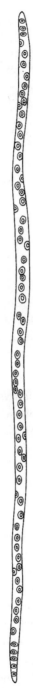

Figure 3.9 Tracheids are narrow water-conducting tubes. The cytoplasm of a tracheid cell dies after the strong walls are laid down. Circular pits are visible on the cell walls.

sap in the spring is a spectacular example of this reversed flow, although the xylem participates as well.

Vascular plants probably appeared on Earth not long after bryophytes, but why did they get so big and their cousins remain so humble? It has sometimes been supposed that bryophytes somehow "failed" to evolve vascular tissue, or more specifically, the strengthening compound lignin that allows tracheids to become stiff enough to withstand the tension of transpiration. After all, some of the larger upright mosses do have the rudiments of a conducting tissue (hydroids), but their cell walls are not rigid enough to form a long-distance conduit. In actuality, the differences between bryophytes and vascular plants, even how tall they can get, has to do more with how they reproduce than with any physical limitations.

Sexual reproduction on land

Stepping onto a water-saturated mat of moss not only disturbs an intricately stored water supply but also possibly an orgy of sexual reproduction. At certain times of the year, those spongy masses are alive with sperm cells swimming about in pursuit of their female counterparts. Early land plants inherited sexual reproduction from the green algae, and the task of the sperm cell is basically the same in both. But in the land plants, there is the added challenge of staying wet. Unlike amphibians, plants cannot waddle or hop back to the pond to release their gametes. They can achieve sexual union only if there is a continuous pathway of water between plants producing sperm and plants producing eggs. The crowded, spongy mats formed by most moss, liverwort, and hornwort colonies not only help maintain moisture for life processes but also provide the liquid highways for sperm movement. Such highways, however, form only for short times after it rains and span relatively short distances.

Like most algae that preceded them, the moss, hornwort, and liverwort colonies we see on the ground or tree trunks are haploid plants. A single set of chromosomes in each cell suffices for the development and functioning of these plants. Because they produce gametes directly, through ordinary cell division, the haploid vegetative phase of bryophytes is referred to as the gametophyte. This is an important difference from animals, which are diploid organisms that produce gametes through meiosis, which you will recall from Chapter 2, reduces the number of chromosome sets from two to one.

As we saw in Chapter 2, green algae usually undergo meiosis in the fertilized egg, which results in zoospores. In *Ulva*, however, we saw that the fertilized egg develops into a multicellular diploid individual, the sporophyte, in which meiosis will later occur to result in many haploid zoospores. Zoospores then grow into another haploid plant that will produce gametes to begin the cycle again. This alternation between gametophytes and sporophytes became the normal situation in land plants.

The gametophyte generation in liverworts, hornworts, and some vascular plants is a flat thallus (see Fig. 3.4), while those of mosses and leafy liverworts are more specialized, with complex leafy shoots (see Figs. 3.2, 3.3, B). In all gametophytes, sperm cells are released in great quantities from chambers called antheridia (Fig. 3.10A), while eggs remain hidden within flask-shaped containers called archegonia (Fig. 3.10B). Pheromone-like chemicals released from the archegonia lure and guide the sperm cells to their destination.

Each archegonium contains just one egg and later serves as a womb for the embryo that develops after fertilization. It is for this reason that the land plants are called embryophytes. The embryo is diploid—the beginning of a sporophyte

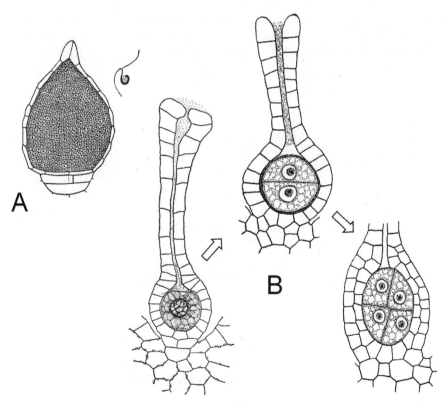

Figure 3.10 A typical bryophyte antheridium (A) consists of a jacket of protective cells and hundreds of flagellate sperm cells. The typical archegonium (B) consists of a vase-shaped jacket with an elongate neck and a single egg sitting at the base. Sperm cells enter through the opening at the top and fertilize the egg. The embryo develops within the chamber until it outgrows it. Drawings from Brown 1935.

phase similar to the one we saw in *Ulva*—but in land plants the sporophyte can be anything from a bag of spores to a seed-bearing tree.

Dispersing genes—the sporophyte generation

Joining of sperm and egg is only part of the reproductive process in plants. It does little to disperse or mix genotypes within or between populations. Plants cannot move around to select a mate, and their sperm cells can move only very short distances, so without any other adaptations, plant reproduction would be largely among branches of the same clone. In most green algae, the zygote divides through meiosis to form motile zoospores, and these are what travel around to mix with genetically different populations. Zoospores are active cells, equipped with chloroplasts for generating fuel and flagella for swimming, and they can seek out suitable sites for settling.

On land, however, zoospores would face the same travel restrictions as sperm cells. Spores of early land plants (and terrestrial fungi) quickly became adapted to survive desiccation and then to be transported by the wind. The adaptive shifts required for this new lifestyle were extensive and occurred in three phases: 1) spores became desiccation resistant; 2) the number of spores per zygote increased; and 3) spore chambers were lifted into the air for better dispersal. These advances were made through the evolution of a separate diploid phase: the sporophyte.

First, the spores had to be completely overhauled to survive the drying effects of air. The flagella disappeared and a hard, waterproof covering based on a stable complex of polymers called sporopollenin developed in the spore wall. The precise chemical structure of sporopollenin has not been fully worked out because it is so difficult to degrade chemically. It is an extremely durable substance that persists in fossilized spores and pollen grains for millions of years. Its presence in spores of fungi, slime molds, and some green algae, as well as in all land plants, suggests that it has evolved independently a number of times.

In *Coleochaete*, sporopollenin is found in the walls of dormant zygotes, which may at times have to survive the drying of their habitat. Most likely, then, the ability to make sporopollenin was already present in the earliest land plants, but its application was shifted from the zygote to the spores. Waterproof walls were no longer needed in the zygote, which was nestled within the womb-like archegonium, but were needed in the spores as they adapted to dry conditions.

Simply stated, the set of genes responsible for development of the desiccation-resistant cell wall, which was previously switched on during zygote development, was switched on instead during spore development. Most likely, this major evolutionary change was brought about by a simple mutation in a master control gene. We will see other examples of this genetically economical kind of evolutionary change shortly.

These newly renovated spores may not have been dispersed by the wind at first, but they were at least suited to sit out dry spells and even survive the death of their parents. They may originally have been produced in simple chambers embedded within the gametophyte tissues, as are the zoospores in *Coleochaete*. They may have been washed to new areas during flooding or blown about with dust and other debris during dry spells, perhaps only after the parent plants had disintegrated. They may also have been transported by animals. But the stage had been set for true wind dispersal.

Spores thus became tiny, dry, dormant, and passive—they could not control where they would land. This was a radical transformation from the swimming zoospores of their ancestors, which could sense and respond to environmental signals in choosing where to settle. Because success of any particular spore landing in the right spot became less likely, the number of spores had to be increased. *Coleochaete* produces from 4 to 32 zoospores after meiosis in each zygote. Each of these zoospores has a fair chance of swimming off and establishing itself in a suitable location, but such modest production of spores would not do for random dispersal by the wind.

The spores of land plants swirl around in the air, potentially being carried for hundreds or even thousands of miles. Where they land is up to chance. If the habitat is suitable, a new colony will be established. If spores from genetically different colonies happen to land in close proximity, a mixed population will result, within which sexual reproduction can take place successfully. The odds are against that outcome, however, so a great many spores must be produced. It is a strategy similar to that taken by red algae (see Chapter 2), fungi, and higher plants with wind-dispersed pollen, such as pines, grasses, and oak trees.

The increase in spore production was accomplished in early land plants by delaying meiosis. Rather than immediately forming a handful of haploid spores, the diploid zygotes of land plants divide through ordinary cell division (mitosis) and form a mass of diploid cells—the sporophyte plant. The sporophyte can get much larger and then undergo meiosis in many cells to produce a large number of spores, all derived from a single zygote.

The formation of separate diploid and haploid bodies is referred to as alternation of generations. All plants, some fungi, and some algae go through such alternation. Animals do not. In plants, it is a necessary adaptation to their landlocked immobility: one generation is optimized to bring sperm and egg together, the next, for dispersing spores as widely as possible.

The development of the sporophyte varies considerably among the four lineages of land plants. In some liverworts, a wall develops around the young embryo, forming a single sporangium that remains embedded within the tissues of the gametophyte (Fig. 3.11). The cells within the sporangium undergo meiosis, resulting in a mass of haploid spores. These spores are obviously not dispersed by the wind, but like the hypothetical ancestors mentioned above, may be transported by floods or on the feet or in the guts animals (Glime 2007).

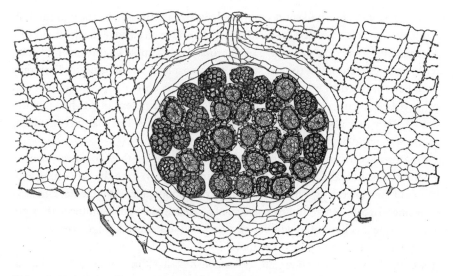

Figure 3.11 In many liverworts, including *Riccia*, spore chambers develop within the tissues of the parent gametophyte. Drawing from Brown 1935.

In most liverworts and mosses, however, a stalk (or seta) develops to lift the sporangium up into the air, just a few centimeters usually, but sufficient to rise above the low mat of gametophytes. At the base of the sporangium, a distinct foot develops that forms a placental connection to the parent plant. Nutrients flow from the parent through this connection, fueling the development of the spores. The diploid sporangium, along with stalk and foot when present, constitute the sporophyte in mosses and liverworts.

The stalk has developed in different ways. In most, it forms as part of the sporophyte (Fig. 3.12A). In some, however, it forms instead from gametophyte tissue (Fig. 3.12B), lifting the entire sporophyte up. In hornworts, no stalk develops, but the sporangium itself elongates, growing from its base over an extended period of time. New spores form in the lower part of the hornwort sporangium as spores higher up are released (Fig. 3.12C).

Note that in all cases, the growth of these stalks and/or sporangia is intercalary ("between"). It comes about through the division and elongation of cells between the sporangium and the parent plant. There is no apical meristem or means of branching, as in ordinary plant growth. This is important, as it contrasts sharply with the sporophytes of vascular plants.

When alternation of generations occurs in multicellular algae, it is for similar but slightly different reasons (see Chapter 2). In *Ulva*, for example, the gamete-producing generation and the spore-producing generation look and function essentially alike, because sperm cells and zoospores travel pretty much the same way. But in land plants, swimming sperm and airborne spores have dramatically different travel requirements, and so the gametophytes and sporophytes that launch them on their respective journeys are quite different. The sporophyte has to not only produce lots of spores, but it also has to get them up into the air. The gametophyte generation, on the other hand, must remain close to the ground, where a watery sperm highway can form.

As an aside, a few bryophytes have found alternate means of dispersing their spores. Capsules of *Splachnum* and other members of the moss family Splachnaceae, for example, remain fleshy and brightly colored, and their spores are released in sticky clumps. *Splachnum* typically grows in or near animal dung and decaying corpses. A musty odor attracts flies that normally feed on the decomposing material toward the spores instead. Flies visiting these moss sporangia become covered in the sticky spores, and when they move off to another pile of dung or another rotting corpse, the spores are rubbed off in a perfect habitat for the next generation (Gerson 1982). This mode of dispersal is rare among lower plants, but strikingly similar relationships with flies have evolved in some fungi and flowers adapted to resemble rotting corpses (see Chapter 7).

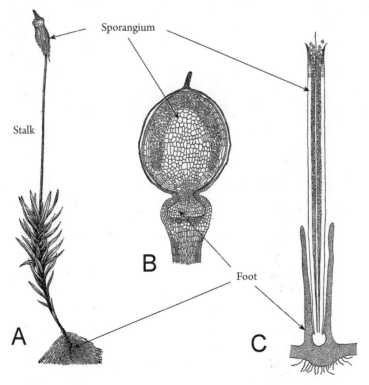

Figure 3.12 A. In the sporophytes of most mosses and liverworts, a stalk develops between the foot and the sporangium, through rapid cell elongation. B. In the sporophytes of *Sphagnum*, the short sporophyte is elevated by part of the gametophyte stem. C. In hornworts, the sporangium elongates at its base. Drawings from Brown 1935 (A) and Haupt 1953 (B, C).

In all nonvascular plants, the sporophytes remain permanently attached to their gametophyte parents, dependent on them for water and nourishment delivered through the placental connection that forms between them. The gametophyte is the dominant vegetative generation, branching to form extensive but low-lying colonies, and it can live for years. The sporophyte is small, unbranched, and ephemeral, existing just long enough to produce a batch of spores and launch them into the air. In the vascular plants, just the opposite is true.

Where did the vascular plants come from?

The gametophytes of vascular plants are even simpler than those of bryophytes. Yet we've already seen that vascular plants are larger and more complex than nonvascular plants. In the vascular plants it is the sporophyte generation that dominates, becoming large, independent, long lived, and able to produce many sporangia over its lifetime. All trees, shrubs, and larger herbs are vascular plant sporophytes.

This fourth lineage of land plants featuring dominant sporophytes is technically referred to as the Polysporangiophyta, for the primary adaptation of their common ancestor was to expand spore production through branching and formation of many sporangia. The earliest members of this group, such as *Aglaophyton* (Fig. 3.13), were small and did not yet have vascular tissues. Xylem and phloem evolved somewhat later, and this is what allowed diploid sporophytes to evolve into larger, more aggressive forms. Plants like *Aglaophyton*, however, successfully exploited the ability for increased spore production and set the stage for the vascular plants.

Ferns are abundant and diverse today, but they represent well the form and life cycle of vascular plants before the advent of seeds. The fern sporophyte is the long-lived generation, living many years and producing a great many spores. The gametophyte, however, is fundamentally similar in structure and function to those of some nonvascular plants, resembling a short-lived liverwort (See Fig. 3.4). Gametophytes are even smaller in seed plants, though still present, hidden within pollen grains and young seeds.

There is still a mystery as to exactly how the vascular plants got started, or more precisely, how their robust, independent sporophytes got started. It was not simply a matter of a moss or liverwort sporophyte detaching itself from its gametophyte and getting taller. The pattern of growth is fundamentally different. The sporophytes of nonvascular plants are simple, unbranched, and produce a single sporangium, while those of the vascular plants are able to both grow and branch by means of apical meristems—dividing cells at the tips of stems.

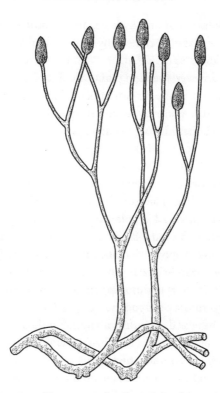

Figure 3.13 A reconstruction of the prevascular plant, *Aglaophyton major*, which consisted of horizontal rhizomes and upright spore-bearing shoots. Redrawn after Kidston & Lang 1921.

It has been suggested that the sporophyte of vascular plants came about as the ability to continue growth and branching evolved in the sporangium stalk of some ancient bryophyte. This is known as the antithetic or interpolation hypothesis of sporophyte evolution. Ligrone et al. (2012) outlined a more sophisticated version of this scenario.

How would the simple stalk of a bryophyte start branching? The stalk that lifts the sporangium of a moss or liverwort is essentially an "afterthought." As we saw earlier, the sporangium of a moss begins its development through division of the zygote into a diploid mass. Later, a few cells between it and the parent plant begin dividing and expanding to produce the stalk. In the stalks of some modern moss sporophytes, cells have the extraordinary ability to expand up to 20 times their original length, but once that's done, there's no way for any new growth or branching to occur. So it would seem that nonvascular sporophytes early evolved into a dead end from which they were developmentally incapable of becoming larger, more complex, or independent.

Another hypothesis has been around for years, however. It proposes that the common ancestor of all land plants had an equal alternation of generations, something like in the sea lettuce, *Ulva*. This is generally referred to the homologous hypothesis of sporophyte origin. From such a beginning, sporophytes in nonvascular plants became simplified and those in the vascular plants became more complex.

By all accounts, however, the land plants are not related to *Ulva* or any other known alga that has an equal alternation of generations. The charophyte cousins of land plants, on the contrary, have no sporophyte generation at all. Their zygotes divide through meiosis to form haploid zoospores or new haploid plants. So the possible existence of equal generations in early land plants is something of a puzzle.

It is possible, however, that both of these hypotheses are partially true. The common ancestor of nonvascular and vascular plants was most likely a haploid plant similar to *Coleochaete* and simple liverworts. It grew and branched indefinitely from apical cells. It formed a simple diploid sporangium through division of the fertilized egg. In its bryophyte descendants, stalks formed but the sporophyte remained simple and dependent.

The branching vascular plant sporophyte may have evolved from this same starting point as the result of a genetic accident. In a bryophyte, the genetic machinery for branching and growth from apical meristems is well developed in the dominant gametophyte generation. The genes for this process are also present in the zygote, but they are normally turned off.

Suppose though, that in some ancient group of land plants, a mutation in a regulatory gene allowed the genes for branching growth to be turned on in the diploid embryo. It would then start growing and branching and would look pretty much like the haploid gametophyte, but would produce sporangia, as it was supposed to, instead of gametes.

At some point then, there may very well have been plants with very similar gametophytes and sporophytes, but through a simple but dramatic genetic change, rather than through descent from algae that had an alternation of equal generations.

Support for this third scenario is coming from genetic research. We know that whole groups of organ-forming genes can be turned on or off by a single regulatory gene. Many examples can be seen in the wealth of knowledge we now have of genetic development in plant models such as *Arabidopsis*. It has also been demonstrated that particular genes perform similar functions in both nonvascular gametophytes and vascular sporophytes. The development of root hairs (slender outgrowths of the root that enhance water absorption) in vascular plant sporophytes, for example, are controlled by the same genes that control the development of rhizoids (threadlike growths that provide anchorage) in moss gametophytes (Menand et al. 2007, Jones & Dolan 2012)). The next few years of genomic studies in both nonvascular and vascular plants may reveal more about the source of the genetic machinery that allowed vascular plant sporophytes to become independent, branching, long-lived plants.

Why are there no moss trees?

Over evolutionary history, many major groups of organisms have had both their giants and their miniatures. Tiny lizards dodged around the feet of dinosaurs, mice steered clear of wooly mammoths, and single-celled planktonic algae still drift among giant kelp. Even among vascular plants, one can find nearly microscopic duckweed in streams running through forests of giant redwoods. Yet nonvascular plants have always remained small. Why?

One possible explanation is that only diploid organisms can get large. Remember that the main vegetative plant of a moss is haploid. Larger organisms may be more vulnerable to damaging mutations during development, so being constructed of diploid cells where detrimental mutations are masked by the normal alleles in the second set of chromosomes may be advantageous (Crow & Kimura 1965). In addition, diploid organisms may have more evolutionary flexibility and evolve faster because they accumulate more duplicate genes or experience higher levels of genetic recombination (Xue et al. 2010). As valid as these factors may be, they may be moot in bryophytes.

Another common explanation for the lack of large mosses is that they, for some reason, "failed" to evolve lignin. This is the resinlike material that penetrates the walls of the water-conducting and supporting cells of vascular plants, making them rigid enough to support extensive upright growth and resist the tension of the water being drawn through them. Lignin essentially made the evolution of large plants possible, and its appearance set the stage for the race to gigantism in vascular plants. Lignin evolved very early in the vascular plants, however, and it would seem that one or more lines of nonvascular plants, in the 450 million years of their existence, would have evolved something comparable, if there were any need for it.

Some mosses do have simple water-conducting tubes, and this has allowed them to become giants by moss standards. Members of the common moss genus *Polytrichum* may grow to a height of 30 cm or more, and the "giant" moss *Dawsonia superba* may exceed 60 cm in total height. This appears to be the upper limit.

So why didn't lignin, or something like it, evolve among these larger mosses and enable them to get taller? In fact, lignins have been found in both the charophyte alga *Coleochaete* and in the hornwort *Anthoceros* (Delwiche, et al. 1989), suggesting that the possibility has always been available to land plants. Evidently, there was no reason—no selective pressure—for nonvascular plants to develop lignin-reinforced tubes, and so this explanation may be moot as well. Something else has kept bryophytes small.

Remember that a moss plant is a gametophyte. Its role in the reproductive cycle is to produce gametes. The sperm cells must swim through a water-logged spongy matrix to reach the eggs of an adjacent, genetically different colony in order to achieve cross-fertilization. The largest mosses must still be packed close enough together so as to hold water between their reproductive tips. The taller they are, the more difficult this becomes. If mosses were trees or even modest shrubs, their sperm cells would be left, quite literally, high and dry—meters away from the nearest egg. Evolution of lignin and vascular tissues in the gametophytes of mosses would have been of little value because of the limit on how big mosses can get and still reproduce sexually.

Vascular plants are sporophytes and face no such restriction. The role of the sporophyte in the reproductive cycle is to produce spores, which are specifically adapted to withstand dry conditions and to be dispersed by the wind. Height is an advantage for this function. Hence vascular plants can become trees while bryophytes can't get any taller than they already are. We will see in the next chapter how early vascular plants developed a variety of growth forms in the competition for light and spore-launching advantage, and how they formed the first great forests.

Figure 4.1 Sequoiadendron trees at maturity are the most massive living organisms on Earth. They are supported by wood laid down in annual rings.

4

Vascular Plants and the Rise of Trees

No one who stands in front of the General Sherman tree (a specimen of *Sequoiodendron giganteum*) (Fig. 4.1) in Sequoia National Park can fail to be impressed. It would take about 20 adults with arms outstretched to encircle the base of this largest single chunk (1487 cubic meters) of biomass in the world. The coastal redwood (*Sequoia sempervirens*) from California currently holds the record for height (115.61 m), though the tallest *Eucalyptus regnans* from Australia is just a meter shorter. The prize for longevity has traditionally been awarded to still another—the bristlecone pine (*Pinus longaeva*), which can live 5000 years or more, though it can be argued that certain clonal plants are even older (as we will see in Chapter 8). The extreme girth, height, and age of these trees are attained by the addition of layers of water-conducting tissue (secondary xylem) each year, to form what we know as wood. Wood not only provides water-transport capacity but also physical support.

These botanical giants are the culmination of the botanical "race to the sky" that began in the Devonian period (419 to 359 million years ago) among the earliest vascular plants. Early free-living sporophytes had the potential to grow and branch indefinitely and could get taller than most nonvascular plants, but they did not yet have wood or even any true vascular tissues. These early plants were only a few centimeters tall, but as the race gained momentum, every small increment in height provided advantage in the competition for light. It also allowed the plants to release their spores a bit higher and hence to be carried further by currents of air. This resulted in an escalating competition that led to taller and taller plants. Within the short 60 million years of the Devonian period, tiny plants with incipient vascular tissues evolved into full-fledged woody trees, their crowns interlocked in the first true forests. For reference, that is about the same amount of time that would later pass between the first primitive primates and the emergence of humans.

The evolution of massive trees with deep root systems began a transition that would change the Earth dramatically. Forest cover accelerated the formation of soil, created a more complex recycling pattern for fresh water, released oxygen into the air, and turned carbon dioxide into biomass. The forest cover also vastly increased the habitat for animal life, and the age of amphibians followed closely.

The removal of carbon dioxide from the atmosphere in the late Devonian by the rapidly expanding cover of vegetation affected the weathering of rock, the accumulation of soil, and the deposition of carbon-rich sediment. This probably contributed to global cooling and even a period of minor glaciation. In addition to cooler temperatures, increased nutrient flow from decaying plant matter may have led to blooms of algae and bacteria that decreased oxygen levels (eutrophication) in shallow seas. These two factors are believed by some to have triggered a biotic crisis in tropical marine environments, where many forms of animal life went extinct (Algeo et al. 2000).

As the early forest resided largely in vast swamps, much of their biomass did not decay but accumulated and turned into coal. The period after the Devonian is called the Carboniferous, in reference to the vast coal deposits formed at that time. The evolution of seed plants toward the end of the Devonian led to a still greater advance in the conquest of dry land. Seed plants transport their sperm cells in pollen grains, rather than through moist soil, and so can live and reproduce in drier conditions. This extended the cover of vegetation up previously barren slopes and into many new habitats, expanding soil development and controlling erosion.

The first vascular plants

The precursors of vascular plants, the early polysporangiophytes, appeared in the mid Silurian, about 425 million years ago. This was some 50 million years after the first evidence of nonvascular plants (Strother et al. 1996). These new kinds of plants probably lived at the margins of wetlands, as they lacked roots or other organs that could penetrate the soil deeply in search of water. They also lacked leaves and so were little more than a system of green forking stems that bore sporangia at their tips. Many of these plants were probably also clonal, spreading along the ground in much the same way as mosses or liverworts.

In some, such as *Aglaophyton* (see Chapter 3, Fig. 13), green stems rambled along the ground, with periodic vertical branches that produced sporangia. In other plants, such as *Horneophyton* (Fig. 4.2), there was a greater distinction between the horizontal rhizomes and the vertical aerial shoots. Where they touched the wet soil, the sprawling stems produced rhizoids, threadlike extensions of individual epidermal cells that provide anchorage and some water absorption. No multicellular roots existed yet. Rhizoids today still provide both anchorage and water absorption for vascular plant gametophytes, but most modern sporophytes have true roots.

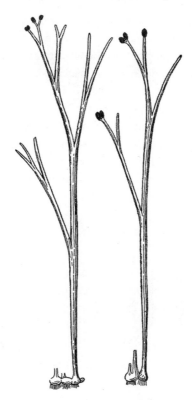

Figure 4.2 Horneophyton lignieri had a compact, well-defined rhizome and forking upright shoots. Drawing from Brown 1935, after Kidston & Lang 1917.

The term shoot generally refers to a young growing stem with whatever leaves, reproductive structures, or other appendages develop from its tip. Rhizomes lie along the ground or under the soil surface, forking occasionally to form extended colonies, and the sporangia-bearing aerial shoots arose from the rhizomes. This clonal model of growth with rhizomes and aerial shoots is fundamental to many kinds of plants, including many advanced flowering plants.

Horneophyton and *Aglaophyton* probably had simple conducting cells similar to those found in some mosses, but not true vascular tissues. Genuine xylem tissue, consisting of tracheids strengthened with lignin, was however present in *Rhynia* and *Cooksonia*, two well-known genera from the early Devonian period. From there, taller, bushier, and more straggling forms emerged.

The earliest vascular plants branched through a more-or-less equal (dichotomous) forking of the stem, and sporangia formed at the tips of some, if not all branches. From that basic model, there was an early split based on different ways to increase spore production. That split is manifest in two ancient and still leafless fossil genera: *Zosterophyllum* and *Psilophyton*. Unequal branching was the key to both strategies.

In *Zosterophyllum* (Fig. 4.3A), the main shoots divided a number of times unequally to produce a series of very short side branches, each ending in a single sporangium, with the main shoot continuing upward barely diminished. A single terminal sporangium was thus replaced by dozens, greatly increasing the spore output.

In *Psilophyton,* (Fig. 4.3B), the shift in branching pattern was more subtle. The main shoots branched unequally to produce slightly smaller side branches, but these side branches themselves forked several times to result in clusters of sporangia. Again, a single sporangium was replaced by dozens. These different branching patterns led to two major lineages of vascular plants: the lycophytes and euphyllophytes, with not only different patterns of sporangia production but also different kinds of leaves and growth forms.

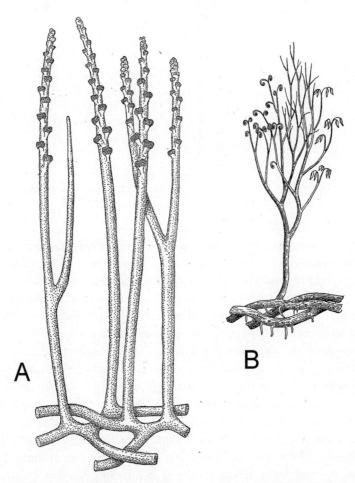

Figure 4.3 In *Zosterophyllum* (A), sporangia were produced on very short branches lined up along the stem tips, while in *Psilophyton* (B), stem tips branched several times to create small clusters of sporangia. Drawings from Smith 1935, after Lang 1927 (A); and Brown 1935, after Dawson 1859 (B).

The origin of roots

Before those new kinds of leaves and growth forms could evolve, better underground systems for gathering water and minerals (i.e., roots) were needed. Early rootless plants probably lived primarily in moist soil near water. True roots allowed the evolution of more complex vascular plants and colonization of drier habitats. Roots most likely began as downward-growing rhizomes, which provided better anchorage for the plants and were able to absorb water from greater depths.

Specialized, downward-growing rhizomes became true roots when they evolved a protective sheath called a root cap and a new way of branching (Fig. 4.4). The root cap is a cone-shaped mass of cells produced by the apical meristem toward the outside of the tip. The replaceable cells of the root cap slough off as the root tips push through abrasive soil. Sometimes a mucilaginous fluid, mucigel, is secreted in the root cap, further lubricating the forward penetration of the roots.

Early rootlike rhizomes branched off of the primary rhizome by forking of the tip, but in true roots, branches emerge from deep within tissues of the older roots or rhizomes, actually pushing their way through the outer tissues to reach the soil (Fig. 4.4B). There is evidence, however, that some roots evolved from leaves rather than from rhizomes, particularly in the lycophytes (see Raven & Edwards 2001 for an extensive review of the origin of roots).

The early rhizomes and earliest roots were often associated with symbiotic fungi called mycorrhizae. In this mutualistic partnership, fungi gather mineral nutrients from the soil and exchange them for carbohydrates from the plants. Such partnerships are still widespread today. The majority of flowering plants have mycorrhizae, which are still more efficient at nutrient gathering than unaided roots.

The first leaves

The variety of growth forms among vascular plants increased dramatically with the evolution of leaves: broad, flat organs adapted for efficient gathering of both light and carbon dioxide. Leaves are for the most part determinate, meaning they expand to a particular size and shape, function for a period of time, and then die. This is in contrast with indeterminate stems and roots that potentially can grow indefinitely.

Leaves evolved in vascular plants at least twice, and in very different ways. The most ancient lineage of leafy vascular plants includes the living club mosses, or lycophytes, which appear to have evolved from an ancestor similar to *Zosterophyllum*, for they inherited the same pattern of sporangia borne along the sides of the stem. The leaves of club mosses are simple, narrow, and supplied by a single strand of vascular tissue running through its length. The simple leaves

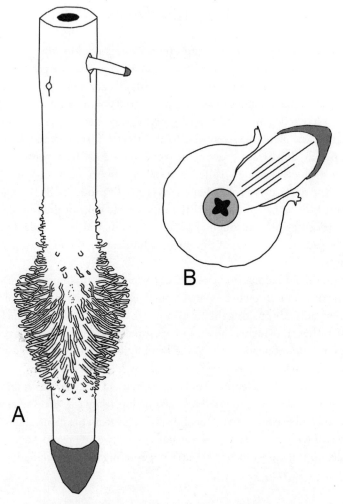

Figure 4.4 A. True roots have a thimble-like covering over the tip, called a root cap, root hairs that help absorb water and minerals, and branch roots emerging from deep within the root. B. A new branch root arises close to the vascular tissues and pushes its way through the outer tissues. Redrawn after Raven et al. 1999.

of lycophytes are called microphylls to distinguish them from the more complex leaves of ferns and seed plants, which are called megaphylls.

Traditionally, microphylls have been thought to have evolved from simple outgrowths of outer stem tissues (Fig. 4.5). In this scenario, small flaps of tissue gradually became longer as a vascular strand extended into them from the main supply in the center of the stem. A more recent hypothesis is that microphylls evolved from the "sterilization" and flattening of sporangia (Kenrick and Crane 1997). Megaphylls, which have probably evolved independently a number of times, evolved from small branch systems that became flattened (Fig. 4.5C).

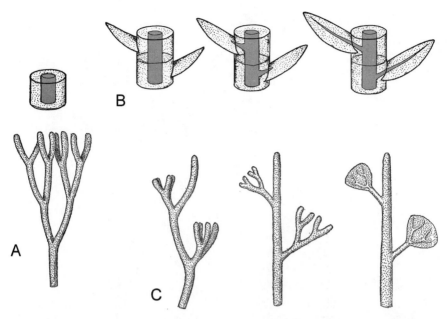

Figure 4.5 From the original leafless common ancestor (A), microphylls are thought to have evolved as outgrowths from the outer tissues of the stem, into which a simple strand of vascular tissue extended (B), while megaphylls evolved through the flattening and coalescence of lateral clusters of branches (C). Drawing from Smith 1938.

Modern club mosses grow slowly, and most have adapted to the low light levels of the forest floor or sometimes to tree branches. They often mingle with true mosses and liverworts. Others are adapted to sunny, wet meadows, where trees and shrubs cannot grow. They resemble mosses in many ways, but they are more robust, as they have vascular tissues and true roots (Fig. 4.6). Sporangia are typically clustered at the ends of aerial stems, each nestled within a leaf or bract (reduced leaf), in some genera forming a distinct cone (strobilus). The genus *Isoetes*, however, is quite different. The plants superficially resemble small clumps of grass or sedge. The narrow leaves arise from a thick condensed stem usually referred to as a corm (as in the corm of a *Gladiolus* or water chestnut). Sporangia are fused to the bases of mature leaves below ground, not in a position for wind dispersal (Fig. 4.7). We'll see in the next chapter how this is part of a rather different kind of reproductive process.

Most plants have larger, more complex leaves referred to generally as megaphylls (also referred to as euphylls, i.e., "true leaves"). The large compound fronds of ferns (Fig. 4.8A), as well as the broad blades of seed plants, are megaphylls, though the two probably evolved separately from leafless precursors. Some botanists have suggested that megaphylls have evolved independently as many as nine times in different groups of ferns and other seedless plants, as well as in seed plants (Tomescu 2008). What does seem to be clear is that all modern plants with megaphylls are descended

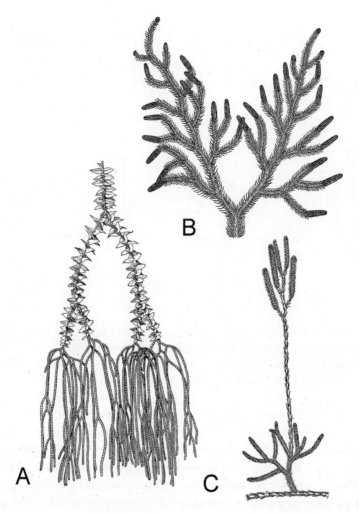

Figure 4.6 Club mosses have short, simple leaves called microphylls, and sporangia that are grouped into cone-like clusters called strobili. Diversity of growth form is represented by A. *Huperzia phlegmaria*; B. *Lycopodiella cernua*; and C. *Lycopodium clavatum*. Drawing from Brown 1935.

from a common ancestor, something like the leafless *Psilophyton* (see Fig. 4.3). The branching side shoots that increased spore production also resulted in more photosynthetic tissue, and some side shoots ultimately flattened to form leaves.

Fern fronds may be considered highly specialized aerial shoots, as they arise from horizontal rhizomes and are both photosynthetic and reproductive. Fern fronds grow rapidly from branching apical and marginal meristems. The branches of the frond unroll from a coiled "fiddlehead" and are fully expanded within a matter of days. Sporangia are borne directly on the underside of fern fronds. The sporangia are typically packed together into discrete clusters called sori

Figure 4.7 Isoetes has very long microphylls attached to a short, stout stem. Sporangia are located in the swollen base of each microphyll. Drawing from Haupt 1953.

(singular: sorus) (Fig. 4.8B, C), and are usually covered by a flap of leaf tissue called an indusium. The varied form and distribution of sori are characteristics by which different kinds of ferns can be recognized. Later, we will see that the first seeds evolved from sporangia on large fernlike fronds.

Ferns emerged as the dominant seedless vascular plants and are still abundant, diverse, and conspicuous today. There are some 12,000 species occupying forest floors, meadows, tree branches, arctic tundra, and rocky alpine slopes. Some even take on the form of trees, vines, or tiny floating aquatic plants (Fig. 4.9).

Cousins of the ferns, called horsetails (sphenophytes), evolved a different form of aerial shoot, one that enabled particularly rapid upward growth. In horsetail shoots, the nodes and internodes are very distinct, and they are involved in a specialized form of growth (Fig. 4.10). Nodes are the points along the stem where leaves are attached, and internodes are the sections of stem between nodes. Any multicellular plant, and even some algae, in which there are distinct leaflike appendages along a slender stem, can be said to have a node/internode type of organization. Buds for new branches may form at the nodes in many plants, and they became standard in flowering plants.

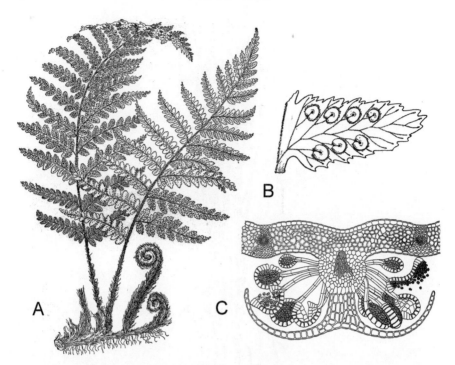

Figure 4.8 A. The upright frond of a fern is a complex megaphyll that bears sporangia on its lower surface; the rhizome extends horizontally. B. Sporangia are typically in clusters called sori that hang from the underside of the fronds. C. Sori are often covered with flaps of protective tissue (lowermost, umbrella-like structure). Drawings from Brown 1935 (A); Haupt 1953 (B), and Transeau et al. 1940 (C).

The leaves of early horsetails, such as *Sphenophyllum*, appear to have been small megaphylls that encircled each node in a whorled arrangement (Fig. 4.11). The first sphenophyte leaves were in fact flat and had forking veins, and they appear to be derived from small branch units of an ancestor similar to *Psilophyton*, most likely smaller subunits than those that gave rise to the complex fronds of ferns (Doyle 2013). In many extinct and all modern horsetails, the leaves were modified into short bracts that protect the soft tissues of the shoots during growth (see Fig. 4.10). Photosynthesis takes place in the main upright stems or in whorls of slender green branches at the nodes.

One genus of horsetail, *Equisetum*, survives to the present, and it gives us insight about the unique structure and development of this group. Horsetail shoots form first as compact buds, with multiple preformed nodes and internodes tightly compressed within. This often occurs in the fall as plants go dormant. At the beginning of the growing season, the shoots grow rapidly as multiple internodes elongate more or less at the same time. The internodes lengthen through cell division and expansion in a band of cells at their base.

Figure 4.9 Salvinia is a tiny floating fern. The apparent roots are actually highly modified leaves, and the roundish objects are sporocarps, which contain sporangia. Drawing from Coulter et al. 1910.

These bands of embryonic cells represent another type of meristem, an intercalary meristem (intercalary means "between," and refers to the fact that these meristems are located between tissues that have already matured). We saw intercalary growth in the stalks of mosses and liverworts and at the base of the sporangia in hornworts (Chapter 3). We will see this kind of growth exploited extensively in the flowering plants as well, especially in the monocots.

The stems of horsetails are also hollow, reducing the amount of tissue that needs to be produced, and this allows them to elongate rapidly. Vascular bundles and fibers are concentrated in the thin rind of the stem, compensating for the hollow centers and providing reinforcing stiffness for the upright shoots. This is remarkably similar to the growth and architecture of modern bamboos. Club mosses and fern fronds, on the other hand, grow primarily through apical meristems.

In horsetails, sporangia are grouped into cones (Fig. 4.12), superficially resembling those of the club mosses. They are more complex, however, having evolved from clusters of forked branches (Fig. 4.12C). One can see a progression in the fossil record, from relatively loose clusters of sporangia to cones with shield-shaped units that fit together to protect in-turned sporangia.

The whisk fern, *Psilotum,* has long intrigued botanists as a possible relic of the early days of vascular plant evolution, for it has neither leaves nor roots (Fig. 4.13A).

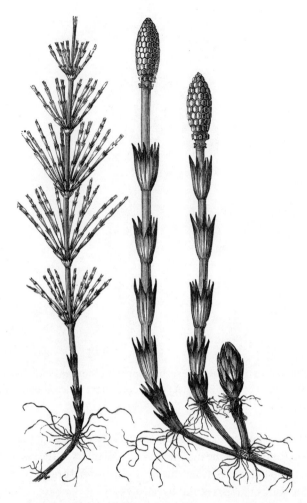

Figure 4.10 Modern species of *Equisetum* consist of jointed, upright shoots arising from underground rhizomes; leaves are reduced to bracts attached to the nodes in a circle; some species have a ring of photosynthetic branches at each node; sporangia are in tightly packed cones. Drawing from Kerner & Oliver 1895.

A close relative, *Tmesipteris* (Fig. 4.13B), has flattened stem sections resembling leaves but no true leaves. Phylogenetic evidence has mounted, however, that places these plants as odd sorts of ferns and not truly archaic plants. The dichotomous branching structure of the aerial shoots does resemble ancient plants like *Rhynia* or *Cooksonia*, but there are tiny flat extensions along the stem that could be reduced leaves. Also, the sporangia are actually compound structures with three chambers (two in *Tmesipteris*), which suggests that more complex sporangia-bearing structures have been condensed.

The most recent evidence reveals that the Psilotales fall into the large clade that includes all the ferns and horsetails (Grewe et al. 2013). The closest living relatives

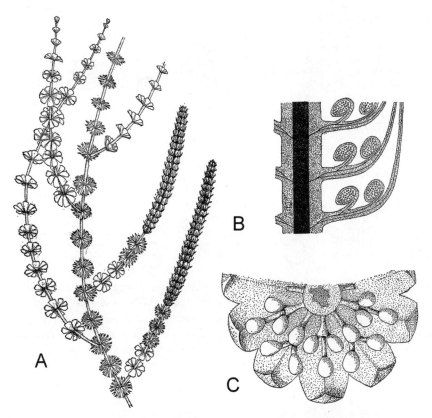

Figure 4.11 An ancient sphenophyte, *Sphenophyllum*, had whorls of fan-shaped leaves at each node. Sporangia were borne in branched clusters on top of scalelike leaves. Drawings from Smith 1938.

of the Psilotales are in the order Ophioglossales (eusporangiate ferns), and both appear to be more closely related to horsetails than to the main group of (leptosporangiate) ferns. *Psilotum* species grow in soil, rock crevices, and in the tropics as epiphytes. *Tmesipteris* is in an epiphyte in the tropical forests of Australia, New Zealand, and New Caledonia. It is not clear why they lack the roots and typical leaves of other ferns, but possibly these features relate to early adaptations as epiphytes subject to frequent dry conditions.

Early trees

Among land plants, we can distinguish two contrasting patterns of growth. In one pattern, the main plant stems (rhizomes) creep through the soil or along a tree trunk, branching occasionally and extending the territory of the plant colony (see Figs. 4.8, 4.10). This is the pattern of typical ferns, club mosses, and horsetails, as well as most bryophytes. Such colonies can live for thousands of years,

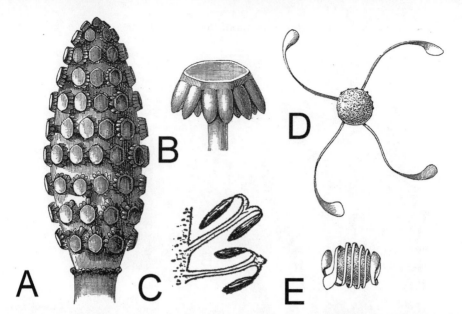

Figure 4.12 The spore-bearing cones (A) of horsetails consist of tightly packed, shield-like units (B) that hide a ring of sporangia underneath. The shield-like units evolved from branched structures (C) in which the sporangia were bent back toward the stalk; spores of horsetails (D, E) have elongate appendages called elaters that twist and turn with changing humidity, pushing the spores out of the sporangia. Drawings from Kerner & Oliver 1895 (A, B, D, E) and Haupt 1953 (C).

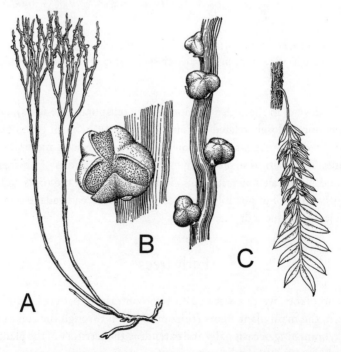

Figure 4.13 *Psilotum* (A) has forking stems and lacks roots and leaves; the sporangia (B) occur in fused clusters of three and are attached along the sides of the stem rather than being solitary at the tip as in the Rhyniophytes; *Tmesipteris* (C) is a cousin of *Psilotum* with leaflike stem extensions. Drawings from Brown 1935.

though they will fragment through decay of older sections. Leaves may be spread out along the horizontal stem individually or clustered in short vertical shoots.

Plants that "move about" by creeping rhizomes have been compared by Halle (2002) to bilaterally (two-sided) symmetrical animals such as centipedes. A fern rhizome does grow at one end (the "head") and decay at the other (the "tail"), producing roots (the "legs") along the way, so moves slowly forward, but the analogy is complicated by the by the fact that rhizomes branch and remain connected to one another.

In the other pattern, exhibited by bird's-nest ferns of the genus *Asplenium* (Fig. 4.14), territorial conquest and immortality are sacrificed for permanent anchorage at a single favorable spot. These plants face upwards, and their leaves form a compact rosette (roundish cluster resembling a rose blossom) around the terminal bud. Halle compares these vertical plants to radially (wheel-like) symmetrical animals such as sea anemones. One could theoretically rotate a radially symmetrical organism around its central axis without affecting its orientation to the environment, but rotating a creeping fern rhizome around its axis would plunge leaves into the ground and lift roots into the air. Bilateral and radial

Figure 4.14 Some ferns, such as this *Asplenium* or bird's-nest fern, have adopted an upright, radial symmetry, and a rosette of leaves.

symmetries are not always as precisely defined in plants as they are in animals, but the analogy is useful for distinguishing the different growth strategies in plants. We will later see radially and bilaterally symmetrical flowers, which are adapted for different pollination strategies.

If the stem of an upward-facing rosette were to gradually increase in height, we would have something we could call a tree. That is exactly what happens in tree ferns, which are a common sight in moist tropical or subtropical forests, particularly in Australia, New Zealand, and other parts of the Pacific region. Tree ferns, which superficially resemble palm trees, have large leaves, divided into many small leaflets and supported by strong cords of fibers in the leaf stalks (petioles). The unbranched trunks of tree ferns consist of a slender core of vascular tissues and fibers, covered from top to bottom with a mass of short, absorptive roots (Fig. 4.15). There is, however, no wood in a tree fern—no way to increase its thickness over time.

The club mosses and horsetails of today survive today as modest, creeping plants, but each had ancient relatives that attained treelike proportions, with trunks thickened with wood. Wood consists of layers of the water-conducting

Figure 4.15 A tree fern, of the genus *Alsophila*, exhibits upright growth of a fibrous, non-woody stem. Drawing from Thomé 1885.

vascular tissue, xylem, which builds up over time. As it builds up, this secondary xylem also provides the physical support necessary to support the increasing weight of the branching crown of the tree.

Wood was made possible by a new invention, the vascular cambium, which is a third type of meristem (after apical and intercalary). The vascular cambium consists of a more-or-less cylindrical array of embryonic cells wrapped around the core of the stem (and in seed plants, of the roots as well), which divide to produce new vascular tissues. In modern trees, the cambium produces new layers of xylem to the inside and new layers of phloem to the outside (Fig. 4.16). The accumulation of wood beneath it causes the vascular cambium to expand over time.

Giant horsetails and club mosses were rather limited as trees go, however, because their vascular cambia could produce only layers of xylem, not phloem. Phloem is the tissue required to transport photosynthetic product from the leaves to other parts of the plant. When the original phloem wore out, the trees declined and died, and so they were probably relatively short lived.

Trees also require a root system that can provide support for a massive trunk and crown. The giant club mosses, such as *Lepidodendron* and *Sigillaria* (Fig. 4.17A,B), sat on pedestals of specialized spreading stems at the base of the tree, and from them small roots, or modified leaves (Raven & Edwards, 2001), emerged to absorb water and nutrients. Giant horsetails, such as *Calamites* (Fig. 4.17C), on the other hand, had massive rhizomes from which their large, bamboo-like upright shoots arose. Like bamboos today, the trunks of these giant horsetails most likely extended rapidly upward through the synchronized elongation of their internodes. Giant club mosses and horsetails formed the basis of the first forests, which appeared toward the end of the Devonian Period. They flourished in extensive swamp forests throughout the Carboniferous Period, building up the massive coal deposits that modern society is now so rapidly turning back into carbon dioxide.

As big as they got, club moss and horsetail trees were short lived. Modern trees live for many years, centuries even, not just because they can continually produce new layers of wood, but because they can also produce new layers of phloem. The key to this was a vascular cambium that could alternately push new cells to the inside and to the outside. Those pushed to the outside became new phloem tissues and those pushed to the inside became xylem. Such a two-faced (bifacial) cambium is characteristic of woody seed plants, but it first appeared in a group of seedless vascular plants called progymnosperms, the most famous of which was the massive *Archaeopteris* (Fig. 4.18). This more versatile cambium arose independently from those in club mosses and horsetails. These stronger, more durable trees came to dominate in the middle to late Devonian period. Their deep roots and leaf litter contributed to the development of the first thick layers of soil and the major ecological changes mentioned at the beginning of the chapter.

The wood of progymnosperms and the early seed plants that followed consisted mostly of the simple tracheids laid down in concentric growth rings. A trunk full

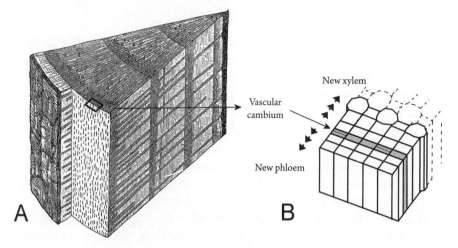

Figure 4.16 In a modern vascular cambium, the cells of the meristematic layer divide so as to produce new cells alternately to the inside (xylem) and to the outside (phloem). Drawing (A) from Ganong 1916.

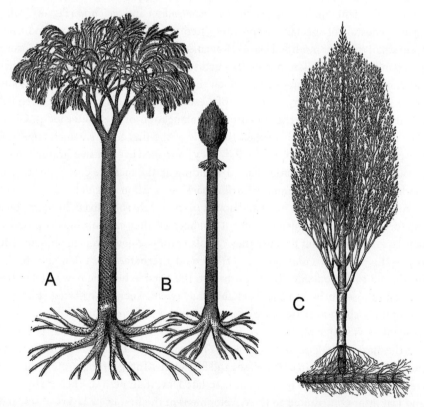

Figure 4.17 Two treelike lycopods from the Coal Age were *Lepidodendron* (A) and *Sigillaria* (B), which had spreading rhizome branches to form a basal pedestal. *Calamites* (C) was apparently more like a bamboo, with upright shoots arising from horizontal rhizomes. Drawings from Smith 1935.

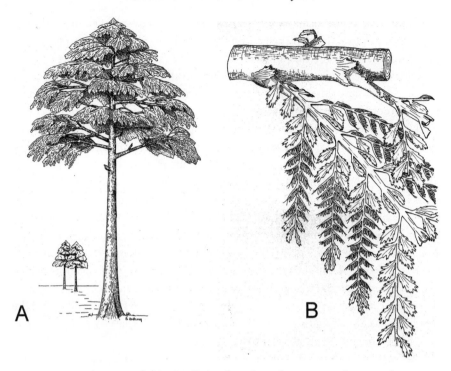

Figure 4.18 Archaeopteris (A) had well-developed woody tissue similar to modern gymnosperms and was the first long-lived tree; some leaf segments were modified to bear sporangia (B). Drawings from Beck 1961.

of densely packed tracheids creates millions of tiny capillary passageways for the upward movement of water. The cohesion of water molecules plus the adhesion of those molecules to the sides of the tracheids allows them to continually move upward as evaporation at the top of the tree drains them away. Thus, movement of water up even the tallest of trees is passive and is due to strictly physical processes. The maximum height of a tall tree represents a balance point between the upward pull of transpiration and the cumulative weight of the water in the tree.

The root-stem axis

Woody plants, possibly including the progymnosperms, made another extraordinary advance. The more robust trunk and massive crown system of a tree required a matching root system to support it. Roots of creeping plants and upright ferns are adventitious, that is, they arise serially from the stem. Such roots disintegrate with time and are replaced by new ones from the younger stem tissues. In erect woody trees and shrubs, a permanent, branching, woody root system, called an axial root system, develops directly beneath the base of the trunk, forming a solid pedestal.

The root-stem axis forms early in the development of the plant embryo. Though we don't know exactly what the embryo looked like in progymnosperms, in seed plants, it is basically rod-shaped, with the shoot apex at one end and the primary root apex at the other end. The root system develops through branching of the primary root. The primary root of the embryo may continue to develop as a dominant taproot, or it may branch into several main roots, each of which becomes thick and woody. An axial root system is also commonly referred to as a taproot system, because even if the primary root does not remain dominant through the life of the plant, the main branch roots all originate from it.

Trees are not perfectly symmetrical because there is randomness in their branching, but from above, they do have an overall roundish shape and radial symmetry, with their branches and roots balanced around the central axis. The trunk of the tree is a relatively narrow connector between the two branched ends, and even a shrub has a narrow "waist" where the stems converge onto the top of the root system. Thus the overall shape of a typical woody plant is like an old-fashioned hourglass (Fig. 4.19). The narrow waist where the stem system meets the root system is fixed permanently to the spot on the ground where the tree began as a germinating seed. This is a key difference between woody plants and the creeping plants that came before them.

The immortality provided by a rhizome system was sacrificed by trees in favor of massive size and height. Trees are committed to a single spot of ground from the time of their birth, but if successful, they can form a massive light-gathering

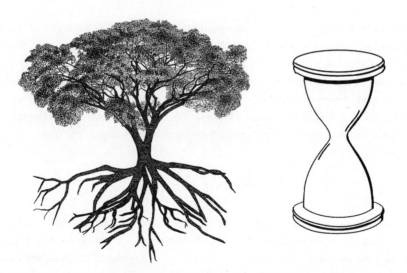

Figure 4.19 Trees have a roughly radial symmetry as they can be rotated around their vertical axis and look the same. A narrow waist connects the leafy shoot system with the branching root system, much as in a traditional hourglass. Redrawn from Brown 1935 and from 4vector.com free clipart.

canopy and dominate that spot for hundreds or even thousands of years. They have the advantage not only of access to abundant light but also to breezes that can carry their spores, pollen or seeds to new locations. Frequent drought, severe winter cold, boggy soil, or persistent strong winds may make it impossible for trees to get tall, and in such areas, shrubs dominate, often scattered among low grasses or other herbs but sometimes developing into elfin forests. A shrub, by definition, is a woody plant with multiple branches from a common root system, a true hourglass shape, with the waist right at ground level.

In these upward-growing plants with permanent, woody crowns, the growing tips of the twigs are continuously exposed to the environment—to freezing cold, desiccating winds, wildfires, and to every kind of animal that would feed upon them—so they have evolved defenses against such threats: bud scales to protect their tender buds during the winter, an endless variety of spines, furry coats, waxes, and toxic compounds to protect their leaves from herbivores, and thick bark to protect their stems from fire. Whether they live for one year or a thousand, trees eventually become old, topple over, and rot away.

The progymnosperms, which gave us the first trees, died out in the early Carboniferous, but somewhere among them the first seeds evolved, and from them, a whole new chapter in the evolution of plants.

Figure 5.1 The evolution of seeds allowed plants to live and reproduce in drier environments. *Welwitschia* occurs in the bone-dry but foggy deserts of southwestern Africa. Drawing from Kerner & Oliver 1895.

5

Seeds and the Gymnosperms

"Mighty oaks from little acorns grow" (anonymous)

A seed, like the spore of fern or moss, is a marvelous miniature vehicle. It contains the genetic instructions for a whole new plant, and it is adapted in one way or another to be dispersed away from its parent into an appropriate location for growth. A seed is far more complex than a spore, however. It contains a multicellular food storage tissue, a partially developed embryo, and a hard, waterproof seed coat. There is a much greater investment in material and energy in a typical seed than in a spore, but the trade-off is in the numerous ways seeds can be adapted for dispersal and subsequent establishment. The evolution of seeds allowed plants to live in a much greater variety of habitats, including even deserts (Fig. 5.1).

The impressive stores of food in acorns and other nuts support embryos that typically germinate in dense shade and have to grow aggressively toward the light. The massive coconut is also a single seed. The white flesh and liquid within serve to nourish the developing embryo as its first roots seek fresh water under salt-saturated beach sand. The coconut, however, is puny next to the largest known seed, the double coconut (*Lodoicea maldivica*), which can weigh up to 22 kg (50 pounds) or more. The exaggerated size of seeds in this forest-dwelling palm is harder to explain.

Some seeds have accessories to aid in their dispersal by animals, wind, or water. Certain pine seeds have wing-like extensions that help them drift in breezes. The seed of the yew (*Taxus*) is surrounded at maturity by a bright-red, edible covering called an aril, which is fed upon by birds. The birds digest the aril and pass the seed with their feces, potentially transporting it many miles from its original location. With a few rare exceptions, spores are unaided and must rely on sheer numbers and the whims of the wind for success.

In their great diversity, some seed plants have reverted back to strategies that emulate spores. Orchid seeds, for example, are as light as dust and a million or more may be produced in a single capsule. They contain virtually no stored food and embryos of just a few cells. This reflects their adaptation to be dispersed through the air by the wind, the same as moss spores. Orchid seeds typically land in drier locations than spores, however, and most are dependent on symbiotic fungi for successful establishment.

So the combination of food reserves, partially developed embryos, a protective seed coat, and dispersal accessories give seeds an advantage in many habitats, particularly upland and drier habitats. However, they did not evolve alone. The "better half" of seed plant reproduction is the pollen grain, which carries sperm cells through the air, directly to the egg developing in the immature seed. This, like internal fertilization in reptiles, was an even more important key to the conquest of dry habitats by plants, for sperm cells no longer had to swim through wet soil to get to the eggs.

Giant horsetails and club mosses persisted in swamp forests through the Carboniferous and Permian but were largely extinct by the end of the Permian period. By the time of the dinosaurs, conifers, ginkgos, cycads, and many seed plants that are now extinct dominated the botanical world. Seedless, spore-dispersed plants survive now in their shadows or in other specialized habitats not yet taken over by the seed plants.

Spores transformed

It might seem then that seed plants did away with spores. On the contrary, spores are still vital parts of the seed-plant life cycle, but their roles have changed. Upon brief reflection, the pollen grains of seed plants are quite like the spores of ferns and other seedless plants, and that is exactly what they are. After dispersal through the air, the role of the typical spore and pollen grain alike is to develop into a haploid gametophyte plant. The gametophyte of the pollen grain is tiny, develops internally, and produces only sperm cells.

The eggs come from a corresponding "female" spore, which is buried within a specialized structure called an ovule. After fertilization, the ovule becomes the seed. Like many other novelties in the evolution of plant life, seeds and pollen grains represent a modification of preexisting structures and processes, which almost always proves more expeditious than creating new organs from scratch.

Among the seedless vascular plants, seedlike structures, that is, hard-walled bodies containing young embryos and stored food reserves, evolved several times, and they can be seen in several unrelated groups of plants still around today. They illustrate the kinds of transformations that likely preceded true seeds.

In the lycophytes *Selaginella* and *Isoetes*, as well as some aquatic ferns, two kinds of spores are produced: large megaspores, which will host the egg-producing side of the sexual process, and smaller microspores, which will provide the sperm cells. This sort of life cycle is termed heterosporous ("with different spores"). Neither kind of spore develops into an independent plant like the gametophyte of a fern;

rather, each subdivides internally to form a miniature gametophyte. In this way, they are similar to the modified spores of seed plants.

As it grows, the multicellular female gametophyte of *Selaginella* ruptures the megaspore wall at the top, where it produces a few archegonia with eggs (Fig. 5.2). Microspores that have landed nearby subdivide within to produce up to 32 sperm cells. When mature, these sperm cells exit the microspore and swim toward the female gametophyte, drawn by chemical signals.

The large female gametophyte consists mostly of food storage tissue, and after fertilization, the embryo develops within the megaspore wall, nourished by these food reserves. This is much the way a seed functions, but in these pseudoseeds, the gametophytes develop in the soil and require fertilization by swimming sperm cells, like other seedless plants. True seeds remain attached to the parent plant and are adapted to receive sperm directly from airborne pollen grains.

The microspores of these heterosporous plants are bulkier than the simple spores of ferns or mosses, because they contain all the food reserves that will be needed to produce a large number of sperm cells. Simpler spores germinate and develop into a photosynthetic plant that will later produce sperm cells. One wonders then how these heavier microspores are dispersed, as they must be to fulfill

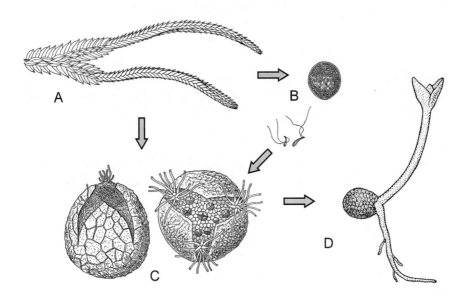

Figure 5.2 Sporangia of *Selaginella* are nestled among bracts that are slightly smaller than ordinary leaves at the ends of leafy stems (A). The male gametophyte (B) also remains within the spore wall, consisting at maturity of a mass of sperm cells that emerge and swim toward the eggs. The female gametophyte (C) develops within the large spore wall. The fertilized egg develops into a diploid sporophyte embryo, which eventually sprouts a shoot and a root (D), much like a seed. Drawings modified from Brown 1935, after Bruchmann (A–C) and Transeau et al. 1940 (D).

the prime directive of sexual reproduction: to mix genetic material from different parts of a population.

Most heterosporous plants, including ferns like *Azolla, Salvinia,* and *Marsilea,* as well as the lycophyte *Isoetes,* are aquatic or semiaquatic, and sporangia are located below water or the soil line. In none are spores regularly exposed to the wind, except perhaps when their habitats dry up and become dusty. It is widely believed that both mega- and microspores of these plants are dispersed via water currents and flooding or by animals such as waterfowl, sometimes as whole sporangia or plant fragments including sporangia (Cox and Hickey 1984).

Selaginella species, on the other hand, mostly live on the ground, on logs, or on tree branches, and their microspores are presumed to be dispersed through the air. However, their sporangia are not elevated on stalks to facilitate wind dispersal, and the heavier spores would not seem to get very far. Some species actively expel their spores, but still they would be flung only locally. There is scarcely any literature on this subject, and so the long-distance dispersal of microspores, particularly of *Selaginella,* remains a mystery.

The pollen grain

Between the microspores of a seedless plant such as *Selaginella* and the pollen grains of seed plants, there is an important trade-off. The relatively heavy microspores, if successfully dispersed, produce a large number of free-swimming sperm cells, each of which has only a slim chance of finding an egg. In wind-pollinated seed plants, however, a great many light pollen grains are produced, because the chance of an individual pollen grain landing on an ovule high up in a tree is extremely small.

Once a pollen grain has landed at just the right spot, however, not so many sperm cells are needed—it only takes one to fertilize the egg. Two are produced in each pollen grain; the second provides a backup, but more would be unnecessary baggage. Thus pollen grains have been adapted for minimum weight and maximum efficiency.

Like the spores of other plants, pollen grains originate through meiosis. Pollen resembles other spores as they are produced in microsporangia (or pollen sacs), but even before they are released from the parent plant, each has subdivided internally to create a tiny two-to-three-celled gametophyte (Fig. 5.3A). Pollen grains are carried by the wind or by animals to the ovules. The "male" gametophyte (microgametophyte) within the pollen grain further develops when it contacts an ovule, producing two sperm cells and a pollen tube (Fig. 5.3B). The pollen tube grows into the ovule and delivers the sperm cells to the egg. The sperm cells of archaic seed plants like ginkgos and cycads are large and equipped with many flagella (Fig. 5.3C), but most have been reduced to little more than nuclei.

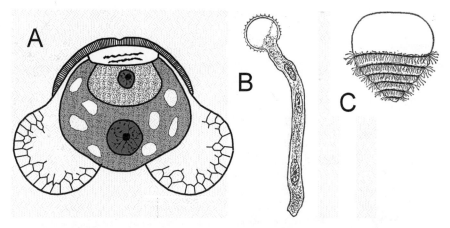

Figure 5.3 The gametophyte within the pollen grain of pine (A) consists of two functional cells and the degenerate remains of two vegetative cells. Upon germination (B), a pollen tube emerges, directed by the pollen tube nucleus at the tip. The generative cell has split to form two sperm cells (from a flowering plant). In *Ginkgo* and cycads (C), sperm cells are flagellate, but in conifers and flowering plants, they are not. Source: Redrawn from Brown 1935 (A, B) and Coulter et al. 1910 (C).

The ovule

Plants like *Selaginella* took a significant step toward the evolution of seeds by creating a protected, food-filled container for the development of their young embryos. Somewhere among the progymnosperms, plants with mega- and microspores made the additional modifications that turned them into ovules and pollen grains. Ovules were retained on the parent plant, sometimes high above the ground, and at the same time, the microspores were adapted for transport through the air.

An ovule is essentially a highly specialized sporangium retained within a protective jacket called an integument. The ovule remains attached to the parent plant until it has matured as a seed. This specialized sporangium, called a nucellus, contains a single small spore surrounded by food storage tissue. This is quite different from the megaspore of *Selaginella*, which is large and itself filled with food. Though small, the "female" spore in the ovule is still referred to as a megaspore, because it ultimately produces eggs, like the large, egg-generating spores of seedless plants. So "mega" and "micro" in seed plants refer to the organs involved in the production of eggs and sperm, respectively.

Meiosis normally results in four nuclei, but inside the ovule, three of the nuclei disintegrate, leaving just one megaspore. That spore never sees the light of day as it develops into a female gametophyte. Food reserves, at first stored in specialized tissues of the nucellus, are transferred to the expanding gametophyte, which eventually produces one or two eggs. In gymnosperms, such as pines and cycads, the rather large eggs are technically still in archegonia, but appear to be directly embedded within gametophyte tissues (Fig. 5.4). Later we'll see that in flowering plants the gametophyte and egg are even simpler.

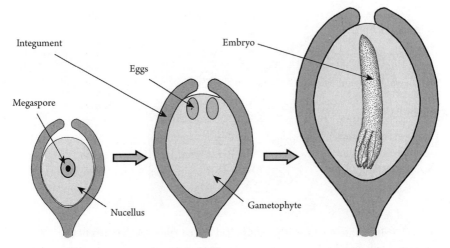

Figure 5.4 The ovule of a pine tree begins as a sporangium, with a single small spore surrounded by stored food. The spore develops into a multicellular, haploid gametophyte with two large eggs, and after fertilization the ovule matures as a seed with an embryo and food stored in the gametophyte tissue.

In gymnosperms, the pollen grain lands directly on the tip of the ovule, and it is drawn by a drop of thick liquid through a small opening into an upper chamber. The pollen tube begins its growth toward the deeply embedded egg, actually digesting its way through the thick nucellus and nourishing itself in the process. This may take weeks or even months. Sometime during the development of the pollen tube, the generative cell divides to form the two sperm cells. In cycads and *Ginkgo*, sperm cells are flagellate (Fig. 5.3C) and motile, even though they do not have far to swim. In conifers and all flowering plants, sperm cells are nonflagellate, and little more than nuclei. The pollen tube carries them directly to the egg.

The seed at this point consists of an outer diploid seed coat, a haploid mass of stored food, and a diploid embryo, reflecting the sexual phases that have taken place within it. When shed from the plant, the seed is more-or-less dormant, depending on the environment, and can be dispersed to new areas by wind, water, or animals. Seed plants thus have two dispersal phases: the pollen and the seed, while spore-dispersing plants have only one. This may be added to the list of advantages held by seed plants.

The seed plant embryo

With the evolution of the seeds, the embryo became more symmetrical. In the last chapter, we saw that woody trees exhibit a roughly radial symmetry and a strong axial root system. At some point, that symmetry became embedded in the embryo. The embryos of seedless vascular plants, such as ferns and club mosses, are initially spherical but then develop into three to four distinct parts (Fig. 5.5A).

Since these embryos remain attached to the gametophyte for some time, the most prominent part is the foot, which serves to absorb nutrients from the gametophyte as well as to serve as an anchor, just as it has since its origin in bryophytes. Smaller bumps represent the future shoot and an embryonic root. In ferns, a fourth bump becomes the first leaf. The first root arises more-or-less opposite the shoot apex, but usually disintegrates early in the plant's life and is replaced by adventitious roots emerging from the developing rhizome.

The embryos of early woody trees became reorganized as the primary root took on greater prominence. The foot region disappeared and the embryo became more-or-less a linear rod (Fig. 5.5B), with growth focused equally on the embryonic shoot and the embryonic root. This is the root-stem axis, which gives rise to the rise to the radially symmetrical, hourglass type of symmetry discussed in the previous chapter.

The first embryonic leaves, or cotyledons, develop as bulges on the sides of the shoot end of the axis and may become the largest parts of the embryo just prior to germination (Fig. 5.5C). They serve to absorb nutrients from the seed and later may emerge above ground and become photosynthetic. As the tree grows, the entire aerial part of the plant develops from the original shoot apex, and the entire root system develops through branching of the embryonic root apex.

The first seed plants

The first seeds appeared among a group of plants known as seed ferns (Fig. 5.6). The early seed ferns had large compound leaves similar to those of ferns, upon which the seeds and pollen sacs were borne directly. The first ovules emerged

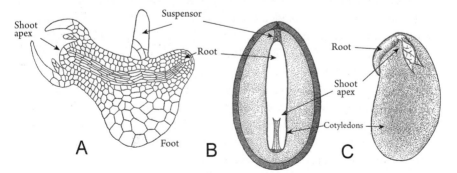

Figure 5.5 The embryos of vascular plants consist of an axis, with a shoot apex at one end and a root at the other. A suspensor pushes the young embryo deep into the nutritive gametophyte tissue. *Selaginella* (A) also has a prominent foot. In pine, (B) the foot is gone. In the bean (C), food reserves have been transferred to the large cotyledons (seed coat removed). Drawings from Haupt 1953 (A, B) and Brown 1935 (C).

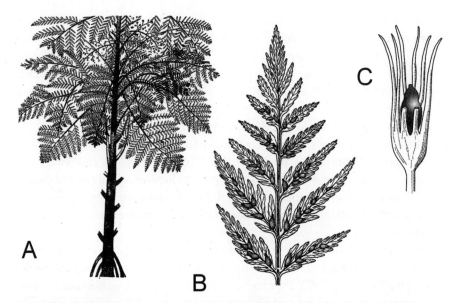

Figure 5.6 Seed ferns were the first true seed plants, and their seeds were borne directly on the large fronds. *Lygenopteris oldhamia* (A) is a well-known early seed fern; *Sphenopteris tenuis* (B) had seeds borne directly on the leaf; *Genomosperma* (C) was an early seed with an incompletely developed integument consisting of a ring of leaf segments. Drawings (A, B) from Brown 1935.

as ancient megasporangia came to be surrounded by leaf segments (Fig. 5.6C), which fit loosely at first, then became fused tightly together, leaving a tiny pore at the top for the entry of a pollen tube.

It is now generally believed that compound leaves evolved independently in ferns and seed ferns from more generalized ancestors, and that the seed ferns emerged from somewhere among the progymnosperms. Most likely, ferns and progymnosperms each descended from ancient seedless vascular plants that had extensively branched photosynthetic stem systems, but which were not yet flattened into blades (Foster and Gifford 1974).

If seed plants were descended from progymnosperms, as is often proposed because of their similar wood structure, they would have been at least somewhat woody from the beginning. However, the earliest known seed plants, such as *Elkinsia* and *Moresnetia*, were fernlike, sometimes vining, and some appeared not to have secondary growth (Prestianni et al. 2007). We must keep in mind, however, the incomplete nature of the fossil record. These may have been specialized plants that inhabited areas where fossilization was more likely, not actually the first plants to have seeds. While we have good fossil evidence for the evolution of seeds in fernlike plants, the connection with the progymnosperms and their form of woody growth remains obscure.

A great many different seed ferns flourished from the late Devonian, through the Carboniferous, to the end of the Permian (around 250 million years ago), mingling with the giant club mosses and horsetails. Those that continued into the Mesozoic were much modified, with seeds and pollen produced on more specialized structures.

Seed plants with smaller leaves

By the late Devonian, about 360 million years ago, seed ferns were flourishing, but a rather different sort of seed plant appeared soon after, in the early Carboniferous period: the widespread order Cordaitales. There were many forms, from large trees in relatively dry forest, to mangrove-like shrubs living in swamps. Exemplified by the genus *Cordaites* (Fig. 5.7), these seed plants were trees with simple, strap-shaped leaves attached spirally around the stems, and strong woody growth like that of *Archaeopteris*. Their seeds were borne in loose aggregations among leaves, or in simple cones, not on fronds as in seed ferns. They were much like modern conifers (pines, redwoods, yews, etc.), which also have simple leaves and seeds in cones, though the precise relationship between these groups is still unsettled.

As mentioned earlier, there is something of a mystery about the relationship between progymnosperms and early seed plants. The intermediate steps of ovule

Figure 5.7 *Cordaites* was an early conifer with simple strap-shaped leaves and reproductive structures arranged in simple cones. Drawing from Haupt 1953.

evolution are found among fossil seed ferns, which had specialized trunks that did not branch much and weak wood development. Cordaites, on the other hand, was a full-sized, freely branching tree with dense wood more like that of the progymnosperm *Archaeopteris*. Because forms similar to *Cordaites* and seed ferns appeared about the same time in the fossil record, some botanists have proposed that seeds evolved independently in the two groups from a seedless progymnosperm ancestor (Chaloner et al. 1977). The generally similar structure of ovules, pollen grains, and the details of the life cycles within all seed plants argue, however, for their common origin.

The simpler, more widely accepted explanation is that seeds evolved in a common progymnosperm ancestor, which then gave rise to the two rather different growth forms: compound fronds in seed ferns and cycads, and the simple leafy twigs of Cordaitales and conifers. In known progymnosperms, the flattened leafy systems resembled fronds, and at least in part were shed as compound units. The growth of the leaf systems, however, was more open ended, like a system of smaller leaf units (see Fig. 4.19). Individual pine needles are therefore probably equivalent to the individual units of a progymnosperm system, while the compound fronds of seed ferns and cycads are equivalent to larger leafy units.

The weaker wood development of seed ferns, or lack of wood altogether in some of them, may be related to their large fronds and sparse branching. This is a common architectural mode found in many unrelated groups of plants, including tree ferns, cycads, and palms. All of these have little or no wood.

The rise of the conifers

During the Permian period, drier conditions, capped by the great mass extinction at the end of that period, completed the demise of the ancient carboniferous forests and seedless trees. Seventy percent of terrestrial vertebrates and a number of orders of insects were lost at that time, creating a very different environment for continued plant evolution. The Cordaitales declined and died out at the end of the Triassic, when the dinosaurs were on the rise. Conifers, cycads, ginkgos, and some specialized descendants of the seed ferns continued into the age of dinosaurs, adapting to more voracious herbivores and drier conditions. Flowering plants most likely arose from a seed fern descendant, one of the so-called Mesozoic seed ferns. During this time seed leaves became variously modified, sometimes condensed into cone-like structures and ultimately into flowers. We will take a closer look at the Mesozoic seed ferns and flowering plants in the next chapter.

Ginkgo biloba is the single surviving species of an ancient lineage characterized by simple, fan-shaped, leaves with forking, or dichotomous, venation (Fig. 5.8). Ginkgos have well-developed wood, like the progymnosperms and conifers. The leaves and reproductive structures are borne on characteristic short shoots on the

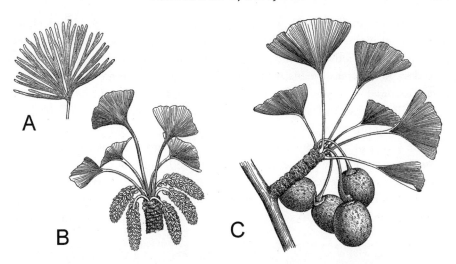

Figure 5.8 The leaf of the ancient ginkgophyte, *Baiera gracilis*, was divided, with forking subdivisions (A). The surviving *Ginkgo biloba* has fan-shaped leaves, with pollen borne on short scales arranged in a catkin (B) and naked ovules borne on short stalks (C). Drawings from Brown 1935.

sides of more elongate branches. Their pollen sacs are borne on simple scales in slender catkins similar to those of the conifers (Fig. 5.8B). The ovules are truly "naked" and borne exposed at the ends of slender stalks (Fig. 5.8C). The ripe seeds resemble fruits, as the outer integument becomes fleshy and fragrant. The fragrance, however, is offensive to most humans, and for that reason male trees are preferred for street plantings. Ginkgos are tolerant of air pollution and resistant to common tree diseases, and so they are valued for cityscapes.

The distinctive characteristics of the ginkgophytes appear to have stabilized in the early Permian period, giving rise to several genera and a number of species over the next several hundred years. The ginkgophyte lineage is probably an early offshoot of the Cordaitales-Conifer clade, though much about seed plant relationships is up in the air. A possible precursor to the ginkgophytes is *Polyspermophyllum* (Archangelsky & Cuneo 1990), which had simple forking leaves and a forking system of ovule-bearing branches.

The seeds of ginkgos may have been dispersed by dinosaurs that found the strong odor attractive, but apparently not so much by mammals. Ginkgos have slowly declined since the Cretaceous and are now probably extinct in the wild. *Ginkgo biloba* exists today only in cultivation, mainly because specimens were preserved in ancient Chinese temple grounds. Buddhist monks have cultivated the ginkgo trees since at least 1100 CE, in part because of the useful properties of the plant. The seed kernels are edible, and extracts of ginkgo are believed to help maintain flexible arteries and brain function in old age.

The conifers are by far the most abundant and widespread of the living gymnosperms. Pine, spruce, and fir trees dominate millions of square miles in the mountains and north woods of Eurasia and North America, as well as the coastal plains of the southeastern United States. They include also the majestic coastal redwoods and massive giant sequoias, as well as yews, cedars, monkey puzzle trees (genus *Araucaria*), *Podocarpus,* and numerous other genera. *Taxodium* is a genus of dominant swamp trees in the southeastern United States. The ample, uniform, fine-grained wood of conifers supplies the bulk of the timber industry.

The leaves of most conifers native to the northern hemisphere are simple, most often in the form of scales or needles (Fig. 5.9), while those occurring in the southern hemisphere tend to have broader, often strap-shaped leaves (Fig. 5.10). The long, slender leaves of conifers, most conspicuously the needles of pines, develop primarily through growth at the base, similar to the growth of internodes

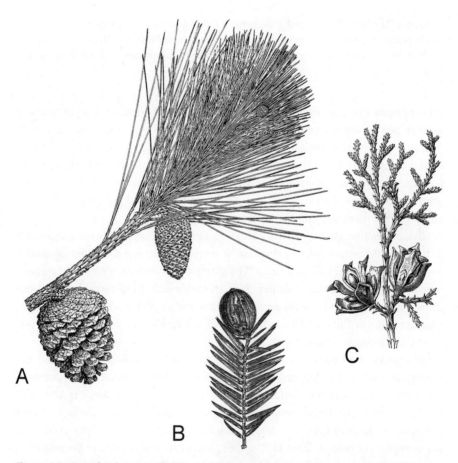

Figure 5.9 Conifers have needlelike, scalelike, or strap-shaped leaves and seeds that are typically in cones (A,C), but sometimes solitary and fruit-like (B). In some, needles are in clusters (A) or twigs (B) that are shed as a unit. Drawings from Brown 1935 (A, B) and Kerner & Oliver 1895 (C).

Figure 5.10 *Agathis* is a conifer from the southern hemisphere, where leaves are often larger than the typical needles and scales of northern hemisphere conifers; seeds are borne within heavy cones. Drawing from Brown 1935.

in *Equisetum* described in the previous chapter. This form of growth is uncommon in leaves, outside of the grasses and their relatives (Chapter 9).

The ovules of conifers are borne in cones that range from massive woody structures to tiny berrylike bodies (see Fig. 5.9B), while pollen-bearing scales form slender catkins similar to those seen in *Ginkgo* (see Fig. 5.8B). Catkins are generally slender and flexible, named for their resemblance to the tails of cats, and pollen is released as they twist in the wind.

The puzzling gnetophytes

The gnetophytes are an enigmatic group consisting of three odd genera that seem to have little in common, and they have long eluded consensus on how they relate to other groups of seed plants. *Gnetum* is a genus of trees and woody vines found in tropical rain forests. Its leaves are remarkably like those of many flowering plants, consisting of a narrow petiole and a broad blade with a netted pattern of veins (Fig. 5.11A). It could easily be mistaken for the leaf of a plum tree or privet. This is a remarkable example of convergent evolution in unrelated plants. Male (Fig. 5.11B, C) and female (Fig. 5.11D, E) reproductive structures are on complex, branched shoots, not in cones. Ripe seeds are exposed along the shoots. The leaves attach to the nodes of the twigs in pairs, opposite one another.

Ephedra is a genus of leafless desert shrubs, found in North America and Asia, with stubby vestigial leaves (bracts) attached to the stem in a circular arrangement (whorls). Its male and female reproductive structures (Fig. 5.12) are in complex structures with small cone-like units.

The third genus, *Welwitschia*, consists of a single species native to the rainless but foggy deserts of coastal Namibia in southern Africa, and it is the most bizarre of all. It consists of a broad stump of a stem sitting on top of a long woody taproot, with long, scraggly leaf segments radiating from the edges (see Fig. 5.1).

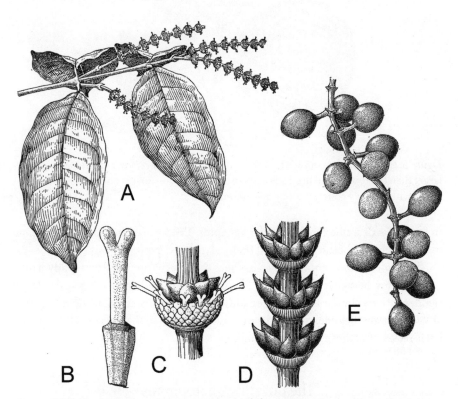

Figure 5.11 *Gnetum* has broad, net-veined leaves (A), similar to those of many flowering plants; its male reproductive units (B) are tiny and grouped, along with sterile ovules, into cupular structures (C) that encircle the slender stems. The young ovules of *Gnetum* are borne in similar structures (D), and they ripen into berrylike seeds (E). Drawings from Brown 1935.

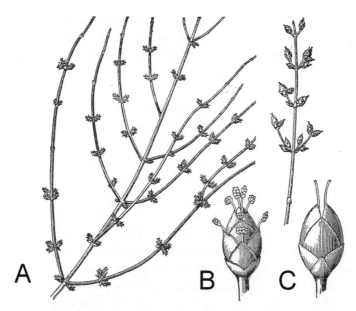

Figure 5.12 *Ephedra* is a genus of desert shrubs with leaves reduced to tiny scales at the nodes of the stems; reproductive structures are housed in compact male (B) and female (C) cones. Drawings from Brown 1935.

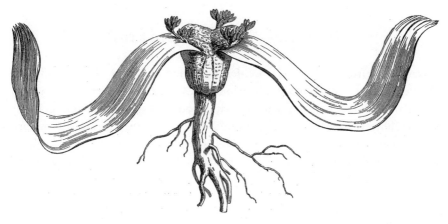

Figure 5.13 *Welwitschia* consists of a stubby stem on top of a long taproot, with two permanent leaves that continually grow from the base; the leaves get wider as the trunk increases in diameter, and they split into numerous parallel ribbons. Drawing from LeMaout & Decaisne 1876.

The stump has no apical meristem. It does not increase in height and does not produce new leaves. The leaf segments around the rim are parts of just two wide leaves (Fig. 5.13) that persist for the life of the plant and split lengthwise into ribbons of irregular width. Like the pine needles mentioned above, these leaves grow from the base, that is, from a basal intercalary meristem, but these leaves grow continuously for the life of the plant, producing fresh new tissues at the base while the older tissues at the tips die and wear off. This is similar to the way leaves grow in grasses and their relatives (see Chapter 9), though there is no relationship. Reproductive structures are housed in compact cones around the rim of the concave top of the stem.

What do these three genera of gnetophytes have in common? One characteristic suggesting relationship among the three is the arrangement of leaves, with two or more attached at each node. There also is some similarity in their branched reproductive axes with numerous bracts. Some have specialized wide water-conducting tubes (vessels), similar to those found in flowering plants, and some have highly reduced female gametophytes, also similar to those found in flowering plants. For these reasons, they have been traditionally grouped together and regarded as the nearest relatives of the flowering plants. Leaves and seed-bearing structures, however, are always simple, suggesting a relationship with conifers and *Ginkgo* rather than with seed ferns and angiosperms, and the specialization of water-conducting cells and female gametophytes are different in fine detail from those of angiosperms. DNA analysis has confirmed the relationship of gnetophytes to one another, but not to the angiosperms. They appear instead to be more closely related to the conifers than to any other known group of seed plants (Chaw et al. 2000).

The cycads—last of the seed ferns?

Cycads date from the late Carboniferous period and were abundant during the Permian and much of the Mesozoic Era. Cycads today are still relatively abundant in some parts of the world, with about 100 species remaining. Many species, however, are rare and endangered, not because they are archaic, but primarily due to overcollection by poachers. These unusual and distinctive plants can command a high price on the black market. Some of the cycads are referred to as sagos, in reference to the starchy pith of the stems that can be dug out and used as a food.

All modern seed plants descended from ancient seed ferns, but cycads have remained the most like them, with large compound leaves, seeds borne on leafy megasporophylls (seed leaves), and sperm cells with flagella. They all are stocky plants that branch sparingly (Fig. 5.14), with weakly developed wood consisting of as much parenchyma tissue as tracheids. Most have simply compound leaves with narrow leaflets, but *Bowenia* and at least one species of *Cycas* (Fig. 5.15) have doubly compound leaves, quite reminiscent of ancient seed ferns.

The seed-bearing structures (megasporophylls) of the genus *Cycas* are conspicuously leaflike, though much simplified compared with ancient seed ferns. The

Figure 5.14 Cycads have stocky unbranched or sparsely branched stems and large compound leaves, superficially resembling palm trees. Drawings from Brown 1935.

round ovules are arranged along the two sides of the seed leaf, below the simplified compound blade (Fig. 5.16A, B). These seed leaves are produced in a whorl around the large apical bud, and after they mature, a new flush of vegetative leaves emerges between them.

In all other cycads, placed in the separate families Zamiaceae and Stangeriaceae, seed leaves are reduced to stubby structures that bear just two inward-turned ovules and are aggregated into compact cones (Fig. 5.16, C, D). The small, thick blades of the seed leaves fit together to form a protective shield for the ovules on the inside. At the time of pollination, they separate slightly, allowing pollen to swirl in or to be brought in by pollinating insects.

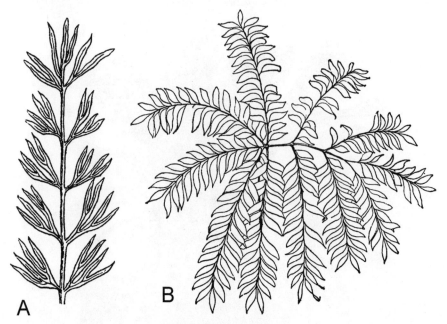

Figure 5.15 Two modern cycads with compound leaflets are *Cycas micholitzii* (A) and *Bowenia spectabilis* (B). Drawings from Seward 1917.

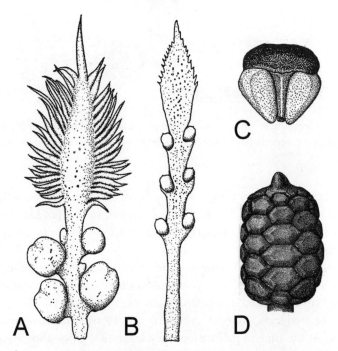

Figure 5.16 The ovule-bearing structures of the genus *Cycas* (A and B) are obviously leaflike and not grouped into cones, while those of other genera, such as *Zamia*, are stubby with shield-like tips (C) that fit together into a compact cone (D). Source: Haupt 1953 (A–C) and Brown 1935 (D).

The primitive flagellate sperm cells, along with the large compound leaves and ovule-bearing structures that sometimes resemble compound leaves, argue that the cycads are the most archaic of living seed plants. Whether or not they could be called "seed ferns" is a matter of semantics. Recent phylogenetic analyses have provided conflicting interpretations of the relationships among living seed plants in general, but particularly with respect to the position of the cycads (Hilton & Bateman 2006, Doyle 2006, Wu et al. 2007, and Lee et al. 2011). Most show cycads emerging from among ancient seed ferns, near the base of the split between other gymnosperms and angiosperms. There is little in the fossil record to show transitions between these major groups, and so their relationships must be considered an open question.

The first pollinators—wind or insects?

It was long assumed that gymnosperms, including the earliest seed plants, were wind pollinated. After all, our most familiar and numerous examples, the conifers, appear to be all wind pollinated, as is *Ginkgo*. We know less about cycads, but a new picture of their reproductive biology has emerged over the past several decades. Some do appear to rely on wind for dispersal of their pollen, but it is now clear that others are indeed insect pollinated. We'll see in the next chapter that the flowering plants (angiosperms) are fundamentally insect pollinated, but under some ecological conditions, they have adapted to other pollen vectors, including the wind. That forces us to rethink pollination in early seed plants. It is likely that many of the Mesozoic seed ferns that preceded angiosperms, some of which had features similar to both cycads and angiosperms, were also insect pollinated.

Pollen is a nutritious food that probably would not have been overlooked for long after the first seed ferns began producing it, and even before that, insects may have been feeding on the spores of seedless vascular plants and bryophytes. Insects appeared on this planet in the Devonian period, at the same time early terrestrial plants were spreading across the landscape. As they are now, most insects then were probably adapted to eat one part of a plant or another, and spores were among the most nutritious parts of plants. So insects were ready to feed on pollen when it came along. In the later Carboniferous period, around 300 million years ago (Bethoux 2009), the first beetles appeared. Beetles today are the largest group of insects, and a great many of them feed on plants. Many feed on pollen. Beetles underwent a great diversification during the Mesozoic, just as seed plants of great variety were proliferating. Wind pollination, as exhibited by conifers, is seemingly wasteful, as you might well think if you've every parked your car under a pollen-shedding pine tree. Pollen is light, dry, and dustlike, and released quickly in large quantities. Much of it escapes being eaten by insects but is largely lost by being blown to the wrong places. Nevertheless, wind pollination works reasonably well in open, breezy areas where the individuals of a particular species form dense stands. Only a few grains out of millions need to succeed for a population of pine trees to sustain its numbers.

Wind pollination does not work so well, however, in thick forests where the vegetation blocks air movement, in places or seasons when there is little or no wind, or in populations in which individual plants are naturally few and far between. Pollen was released from exposed sporangia on the branches or fronds of early seed plants with little that could be done to protect it, so much of it may have been eaten. It is likely that at least some early seed plants evolved to take advantage of the predictable movements of thieving pollen eaters. By making pollen stickier so that it did not blow away in the wind, it became more vulnerable to being eaten, but it also was likely that some of it would get stuck out of reach on the insects' bodies and get rubbed off when they walked across young ovules. Such a strategy fed the thieves sufficiently to reward and reinforce their visiting behavior, while insuring that enough grains would get carried to the ovules of other plants. Wind pollination and pollination by pollen eaters both seem wasteful, but each is successful in the appropriate environment, making the expense worthwhile.

Living cycads, which have changed little in the last several hundred million years, give us some clues as to what beetle pollination may have been like in the dinosaur age (Norstog 1987). Cycads produce pollen and ovules in large cones separately on "male" and "female" plants. This separation is effective protection against self-fertilization, but creates a challenge in getting pollen from one plant to another. Though some cycads are pollinated by the wind, many lure beetle pollinators in with distinctive fragrances and colors.

In some, beetles feed directly on pollen, but in others an alternate food is provided in tissues beneath the pollen sacs, preserving the pollen for its reproductive function. In some cycads, the beetles lay their eggs in this food-rich tissue of the male cones, and the emergence of the young adults coincides with the release of pollen. In any case, the beetles eventually run out of food in the cone and depart, invariably with pollen grains stuck randomly to their underbellies, legs, and other body parts that they can't reach.

The female cones typically lack anything nutritious, outside of the well-protected ovules, but they often mimic the male cones in color and smell. Responding to these cues, the beetles visit the female cones briefly, essentially by accident. Once they discover that they've been duped, they leave. That doesn't take long, but it's long enough for pollen grains to rub off on the ovules. The process seems to be quite effective, generally resulting in successful pollination of more than 90% of the ovules (Stevenson et al. 1998). When no male cones are available, the beetles may burrow into the soil and await the next reproductive season.

Early flowers probably drew upon the same pool of beetle pollinators, but with a radically different design of their reproductive structures. We take them up in our next chapter.

Figure 6.1 The dry, rugged, and varied terrain of South Africa supports a rich flora of spring flowering herbs, succulents, and hardy shrubs, and may represent the type of environment in which the first angiosperms evolved.

6

Darwin's Abominable Mystery

"The rapid development as far as we can judge of all the higher plants within recent geological times is an abominable mystery."

Charles Darwin.

When Charles Darwin made this famous remark in a letter to his colleague J. D. Hooker, in 1879, he was referring to the flowering plants, or angiosperms. As far as could be seen from the fossil record, this vast group of plants burst onto the scene rather suddenly, fully recognizable and in diverse familiar forms. There was virtually nothing to indicate where the flowering plants came from, or any intermediate forms to tell us how their unique reproductive structures came into being. That picture has improved over the past century, but there are still many gaps to explain.

The gaps may be due in part to a long residence by preangiosperms and early angiosperms in habitats where fossilization was unlikely. The exact nature of such habitats is still controversial, but a semi-dry rugged terrain, such as is found today at the southern end of Africa, presents one possibility. Not only is fossilization rare in such a habitat, but several factors here also promote rapid evolutionary change.

Rain falls mainly in the winter here, as it does in southern Europe and California, creating a Mediterranean type of climate with a relatively short growing season. Growth and reproduction must be done in a hurry in a highly competitive environment, and then measures must be taken to survive the long dry season. These upland habitats also tend to be fragmented by rocky terrain into patches of varying moisture, temperature, and soil conditions. This breaks up populations of plants into small localized units, which leads to isolation and rapid evolutionary change.

The moistening of the soil in the winter releases a frenzy of growth and reproduction in plants that have been dormant for nine to eleven months, blanketing the usually barren fields and rocky hillsides with brilliantly colored wildflowers (see Fig. 6.1). A few weeks later it all disappears as quickly as it came. The garish display put on by the flowering plants serves to draw attention to their reproductive organs. Insects, birds, and occasionally other animals, also urgently attempting to feed and reproduce during the short season, descend upon the cacophony of color in search of nutritious pollen, nectar, and other edible products. As those hungry visitors make their rounds, they inadvertently carry pollen grains from one flower to another, achieving pollination.

Angiosperms are distinguished from other living seed plants by unique sets of adaptations that make the reproductive process quicker and more efficient, that exploit animals in a great variety of ways for both pollination and seed dispersal, and that enable a great flexibility of growth forms, including many ways to sit out the dry season in a dormant state. The word "angiosperm" means "hidden seeds," referring to the fact that ovules develop into seeds within unique closed chambers called carpels.

The native plants of South Africa and similar regions around the world today are among the most progressive and diverse of angiosperms, and include members of the legume, sunflower, euphorbia, iris, and grass families, to name only a few. These families represent the current cutting edge of plant evolution. Between 140 and 200 million years ago, the cutting edge consisted of the early flowering plants.

Botanists such as Daniel Axelrod (1952) and G. Ledyard Stebbins (1974) proposed that semiarid subtropical uplands, like those of South Africa, serve as "cradles" of evolutionary innovation, where successive waves of new plant forms appear, diversify, and then are replaced by still newer forms. If this model of evolutionary cradles is correct, it helps to explain Darwin's "abominable mystery." If early angiosperms and their precursors lived in hilly, semidry environments, where fossilization rarely takes place, they would not have left any traces in the rocks.

Flowering plants, and the seed plants leading up to them, may have lived in upland environments for millions of years before some of their descendants moved into the forests and swamps of the lowland flood plains, where fossilization was more likely. The record of angiosperm evolution began only then, and by that time there were already many different kinds of flowering plants.

The most archaic angiosperms alive today, those retaining some characteristics of the ancestral stock, live mostly in isolated tropical forests, and so it has long been inferred by many that this was where angiosperms began. To Stebbins, however, such forests are "museum" habitats that harbor relicts from earlier waves of evolution that began in the uplands. Lowland moist forests would have provided no incentives to shorten the life cycle or invent the new forms of vegetation so characteristic of angiosperms.

Stebbins' model of a dry upland birthplace for the angiosperms has been challenged in a number of ways. Some have proposed that the first angiosperms were aquatic plants (the "wet and wild" hypothesis), because the water lily order, Nymphaeales, is one of the earliest branches of the angiosperm tree. Some early fossil angiosperms like *Archaefructus* appear also to have been aquatic (see Feild et al. 2004 for an extensive discussion of the various hypotheses).

Taylor Feild and colleagues (2004) provided still another hypothesis. They examined the physiology of living archaic angiosperms, that is, relicts of the earliest branches of the angiosperm family tree, and found them fundamentally adapted to moist, shady, disturbed habitats. Moreover, these adaptations appear to have a common genetic basis inherited from a common ancestor, suggesting that they stemmed from the earliest angiosperms. According to this "dark and disturbed" hypothesis, clearings created by treefalls, river fluctuations, storms, or other disturbances were the driving force for the shortening of the life cycle and vegetative flexibility inherent to angiosperms.

Are these hypotheses mutually exclusive? Like blind men touching different parts of an elephant and coming up with different descriptions of it, we may be looking at different phases of the angiosperm story. The implication of the dark and disturbed hypothesis is that all flowering plants alive today (what we call the angiosperm "crown group") had a common ancestor that was adapted to those damp conditions. Whether that is true or not remains to be verified, but one thing we have to keep in mind is that the common ancestor of the crown group was not likely the very first angiosperm.

The crown group ancestor, which may have lived in a dark, disturbed environment, had ancestors and cousins that would also qualify as angiosperms but have left no living descendants. The distinctive characteristics of angiosperms evolved most likely over a very long period of time among those ancestors. The ancestors and cousins together are referred to as the angiosperm "stem group" (Fig. 6.2), and it would be surprising if this assortment did not live in a variety of habitats, including dry uplands.

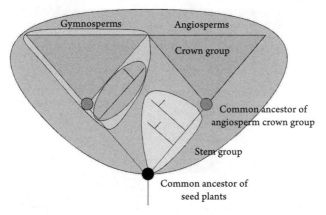

Figure 6.2 The angiosperm crown group includes all living angiosperms and their common ancestor. Earlier angiosperms, preangiosperms, and closely related Mesozoic seed ferns belong to the angiosperm stem group.

Though he did not use the term, I believe Stebbins' scenario for an upland origin of angiosperms applies to the stem group rather than the crown group ancestor. Basic features of the flower and shortening of the life cycle could very well have evolved in the dry uplands and then been followed by refinements added in the dark and disturbed forests—a one-two punch leading to the angiosperms we see today. A third punch, if you like, led some early angiosperms into even damper habitats, giving us the water lilies and their relatives.

The bisexual flower promotes efficient pollination

So how would we recognize the first angiosperms if we saw them? The most obvious feature is the flower itself, which in its most common form is a bisexual structure that brings male and female reproductive organs into close proximity. Doing so was presumably an adaptation for more efficient pollination and one of the great innovations of the angiosperms.

Though there is much variation among living angiosperm flowers, it appears that all are based on a common model, with three sets of distinctive organs: tepals, stamens, and carpels (Fig. 6.3). The pollen-bearing organs, the stamens, lie to the

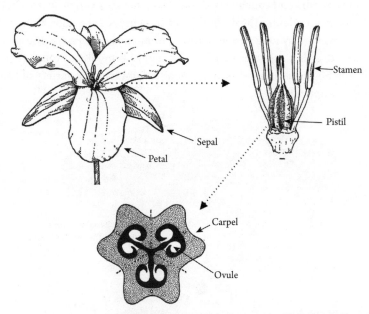

Figure 6.3 The parts of a standard angiosperm flower, as exemplified by the genus *Trillium*. The tepals are specialized into an outer leaflike set, the sepals, and an inner attractive set, the petals. The six stamens are followed by three carpels joined into a common pistil, with two rows of ovules in each. This brings the three stigmas (at the tip of the pistil) together and enables them to be pollinated by a single visitor. Drawings from Haupt 1953.

inside of the tepals, and the ovule-bearing organs, the carpels, occupy the center of the flower.

To achieve pollination, early angiosperms most likely drew upon the existing populations of beetles and other primitive insects that fed in the cones of cycads and other Mesozoic seed plants (see Chapter 5). The cycad "strategy" was to trick the beetles into visiting female cones when they were actually searching for fresh, nutrient-rich male cones. The strategy of angiosperms was to exploit the same insect behavior but within individual flowers. In archaic angiosperms today, arriving insects typically encounter carpels first, where they rub off any clinging pollen grains from previous flowers. They then feed within the flower for a period of time before departing for another flower.

The relatively archaic flowers of *Magnolia* and water lilies (*Nymphaea*) give us two examples of how early bisexual flowers may have controlled the movement of insects and pollen. The numerous carpels in a *Magnolia* flower are attached to an elongate central tower (Fig. 6.4) that serves as a convenient landing place for beetles who have been lured in by the musty fragrance, generated heat, and mild colors of the petals (Azuma et al. 1999). As the arriving beetles crawl down the column of carpels, pollen from previously visited flowers is rubbed off on the sticky carpel tips (stigmas). When the beetles get to the bottom of the flower where the

Figure 6.4 In the archaic flower of *Magnolia*, numerous separate carpels occupy a central tower where beetles land; the long, sticky stigmas of the carpels remove pollen grains from the bodies of the beetles, who then proceed to the bottom of the flower to feed on fresh pollen.

stamens are, they feed until the pollen runs out and then fly off to another flower to start the cycle again.

In water lilies, instead of being located on a tower, the carpels are sunk into a chamber. Insects are lured into it as the flowers open, but there is no pollen available yet. The flowers close at nighttime, trapping the insects until morning, and only then is the pollen released. The beetles feed for a while, and then fly off with a few grains stuck to their bodies. Thus both physical arrangement and timing have been used to manipulate the movements of pollinating insects since the early days of angiosperms, and before that, in cycads and other Mesozoic seed plants.

Animal pollinators tend to forage on one particular kind of flower at a time, thus carrying pollen efficiently between flowers of the same species, and little pollen is wasted. Flowers, in turn, have evolved specific cues and rewards for the animals specialized for feeding in them, reinforcing the loyalty of the animals. The symbiotic relationship between flower and pollinator was so lucrative for both that new kinds of flowers and new kinds of flower-feeding animals proliferated, in an escalating seesaw of coevolution. Wind- and water-pollinated flowers have also evolved. They are frequently unisexual, with inconspicuous male and female flowers sometimes clustered separately into compact structures resembling cones. This has expanded even further the vast diversity of angiosperms.

Both water lilies and magnolias are probably specialized in a number of ways, but the general consensus is that the earliest angiosperm flowers were bisexual. Similar flowerlike arrangement of reproductive organs have been found among gymnosperms (such as the extinct Bennettitales discussed below), but they are different in important details and so are considered different in origin. The details include unique and distinctive structures of stamens, carpels, and ovules.

The floral organs

Both the stamens and the carpels are of a unique structure, distinct from the pollen- and ovule-bearing organs of any other seed plants. The ovules are also highly distinctive, with several unique adaptations that can be tied to the shortening of the life cycle, and some other peculiarities with no obvious adaptive value. These highly distinctive reproductive organs are fundamentally the same throughout the angiosperm crown group, so they were probably standardized prior to the common ancestor.

In addition, carpels, as closed ovule-containing chambers, most likely evolved before they became part of the bisexual flower. In some of the oldest angiosperm fossils, such as *Archaefructus*, the carpels and stamens appear to be located on different parts of branching systems, without evident sepals or petals (Sun et al. 2002), though this structure can also be interpreted as a specialized inflorescence of small flowers (Friis et al. 2003). In any case, it is possible that "angiosperms" evolved before "flowering plants," and that these two names are not exactly synonymous.

Tepals (or the perianth) are the more-or-less leaflike organs that surround the reproductive organs in bud and typically spread out when the flower opens, forming a display to attract animal pollinators. Archaic angiosperms mostly have spirally arranged tepals that sometimes grade from outer, green, protective units into inner colored units. In most modern flowers tepals are divided into two specialized sets: inconspicuous green sepals and showy petals. The sepals are usually specialized for protection of the unopened flower buds, while the typical petals are colored and sometimes also produce fragrances and nectar, as required to attract particular kinds of pollinators. The numerous exceptions to these generalities will be explored in Chapter 7.

Tepals do not necessarily have a single common origin. Any nearby leaves can be recruited either for protection or for display, and this has happened multiple times in the evolution of modern angiosperms. A most extravagant example is found in the familiar poinsettia. The bright red structures that attract pollinators are actually colored leaves that surround clusters of tiny, inconspicuous flowers. It appears on the other hand that petals in some families of flowering plants (e.g., the carnation family and rose family) evolved from the modification of some of the stamens rather than from ancient tepals.

The stamens typically consist of four parallel pollen sacs, usually packed like a bundle of sausages into a compact anther (Fig. 6.5A) and attached to the flower axis by a long, slender stalk (the filament). The four pollen sacs are distinctively borne on two sides of a flattened central ribe (Fig. 6.5B), suggesting an origin from a flat blade. Anthers in several archaic families, moreover, are more blatantly flat and leaflike (Fig. 6.5B–E), as if they had evolved directly from earlier pollen-bearing seed fern fronds. This interpretation is controversial, however, and we'll see shortly that there are some other possibilities.

The carpels are specialized chambers within which the ovules develop and mature into seeds. The angiosperms get their name ("hidden seeds") from this unique arrangement. Pollen tubes must enter the carpel to bring sperm to egg. A specialized region at the tip of the carpel, called the stigma, becomes sticky when eggs are ready for fertilization. This not only catches the pollen, but also secretes chemicals that stimulate the pollen tube to begin its growth. An elongate section of the carpel, called the style, provides a soft-tissued conduit for the pollen tubes to reach the ovule chamber, or ovary, below.

The classic model of the carpel is something like a modern peapod, or the similar multiseeded follicles seen in some members of the Ranunculaceae. Such a carpel is considered a seed leaf (megasporophyll), folded along its midrib (or "backbone"), with the margins joined together and the ovules brought inside (Fig. 6.6). Though there is some controversy about the form of the very earliest carpels (which we'll get to later), most modern carpels can be viewed as

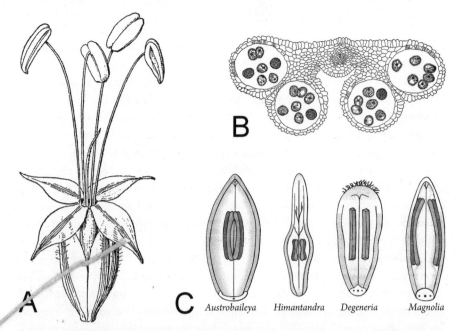

Figure 6.5 Although there is much variation, typical stamens, like those in *Plantago* (A), consist of a slender stalk, or filament, and a slightly flattened anther with four pollen sacs (B). In some archaic angiosperms (C), stamens are flat and leaflike, suggesting ancestry from a pollen-bearing leaf. Drawings from Brown 1935 (A, B) and Mauseth 2014 (C).

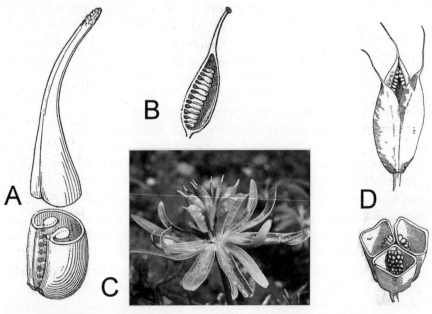

Figure 6.6 A follicle (A, B) is a simple carpel with edges rolled together and sealed along a suture. The ovules are lined up along each margin of the leaflike ancestral structure. In *Eranthis* (Ranunculaceae) (C), the separate carpels are maturing into follicles; petals and stamens have fallen off. The capsule of a *Colchicum* (lily order) (D) develops from carpels fused into a compound pistil. Drawings from Gray 1879 (A, B) and Thomé 1877 (D).

inside of the tepals, and the ovule-bearing organs, the carpels, occupy the center of the flower.

To achieve pollination, early angiosperms most likely drew upon the existing populations of beetles and other primitive insects that fed in the cones of cycads and other Mesozoic seed plants (see Chapter 5). The cycad "strategy" was to trick the beetles into visiting female cones when they were actually searching for fresh, nutrient-rich male cones. The strategy of angiosperms was to exploit the same insect behavior but within individual flowers. In archaic angiosperms today, arriving insects typically encounter carpels first, where they rub off any clinging pollen grains from previous flowers. They then feed within the flower for a period of time before departing for another flower.

The relatively archaic flowers of *Magnolia* and water lilies (*Nymphaea*) give us two examples of how early bisexual flowers may have controlled the movement of insects and pollen. The numerous carpels in a *Magnolia* flower are attached to an elongate central tower (Fig. 6.4) that serves as a convenient landing place for beetles who have been lured in by the musty fragrance, generated heat, and mild colors of the petals (Azuma et al. 1999). As the arriving beetles crawl down the column of carpels, pollen from previously visited flowers is rubbed off on the sticky carpel tips (stigmas). When the beetles get to the bottom of the flower where the

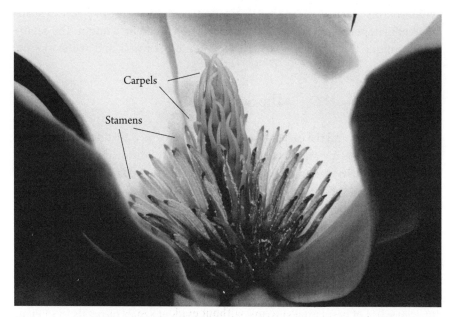

Figure 6.4 In the archaic flower of *Magnolia*, numerous separate carpels occupy a central tower where beetles land; the long, sticky stigmas of the carpels remove pollen grains from the bodies of the beetles, who then proceed to the bottom of the flower to feed on fresh pollen.

stamens are, they feed until the pollen runs out and then fly off to another flower to start the cycle again.

In water lilies, instead of being located on a tower, the carpels are sunk into a chamber. Insects are lured into it as the flowers open, but there is no pollen available yet. The flowers close at nighttime, trapping the insects until morning, and only then is the pollen released. The beetles feed for a while, and then fly off with a few grains stuck to their bodies. Thus both physical arrangement and timing have been used to manipulate the movements of pollinating insects since the early days of angiosperms, and before that, in cycads and other Mesozoic seed plants.

Animal pollinators tend to forage on one particular kind of flower at a time, thus carrying pollen efficiently between flowers of the same species, and little pollen is wasted. Flowers, in turn, have evolved specific cues and rewards for the animals specialized for feeding in them, reinforcing the loyalty of the animals. The symbiotic relationship between flower and pollinator was so lucrative for both that new kinds of flowers and new kinds of flower-feeding animals proliferated, in an escalating seesaw of coevolution. Wind- and water-pollinated flowers have also evolved. They are frequently unisexual, with inconspicuous male and female flowers sometimes clustered separately into compact structures resembling cones. This has expanded even further the vast diversity of angiosperms.

Both water lilies and magnolias are probably specialized in a number of ways, but the general consensus is that the earliest angiosperm flowers were bisexual. Similar flowerlike arrangement of reproductive organs have been found among gymnosperms (such as the extinct Bennettitales discussed below), but they are different in important details and so are considered different in origin. The details include unique and distinctive structures of stamens, carpels, and ovules.

The floral organs

Both the stamens and the carpels are of a unique structure, distinct from the pollen- and ovule-bearing organs of any other seed plants. The ovules are also highly distinctive, with several unique adaptations that can be tied to the shortening of the life cycle, and some other peculiarities with no obvious adaptive value. These highly distinctive reproductive organs are fundamentally the same throughout the angiosperm crown group, so they were probably standardized prior to the common ancestor.

In addition, carpels, as closed ovule-containing chambers, most likely evolved before they became part of the bisexual flower. In some of the oldest angiosperm fossils, such as *Archaefructus*, the carpels and stamens appear to be located on different parts of branching systems, without evident sepals or petals (Sun et al. 2002), though this structure can also be interpreted as a specialized inflorescence of small flowers (Friis et al. 2003). In any case, it is possible that "angiosperms" modifications of this fundamental design. Many carpels become dry and split open to release the mature seeds. The splitting occurs most often along the joined margins, essentially "unfolding" the original leaf, though many other ways of opening have evolved among specialized fruit types.

In archaic angiosperms, carpels are separate from each other and often variable in number. In most modern angiosperms, however, the carpels are fused together, creating a multichambered pistil. An advantage of this arrangement is that several carpels then share a common stigma, or have their stigmas close together. That in turn results in quicker and more efficient visits by pollinators. The pistil or the individual carpels become fruits as the ovules within them ripen into seeds.

As the angiosperms diversified, carpels became both simpler and more complex. Some may harbor a single ovule (as in a plum), or at the other extreme, a million or more (as in an orchid pod). In many angiosperms when the carpels are fused together into a pistil, the chambers of the carpels remain distinct from one another, and in each chamber, the two rows of ovules associated with the ancient carpel edges are lined up along a central axis. The capsule of a lily illustrates this nicely (Fig. 6.6D). In others, the carpels join edge to edge, creating a single chamber.

Hiding ovules within closed carpels contributed to the early angiosperms' ability to grow and reproduce in varied climates, particularly dry ones. It protected young ovules from drying out or being eaten by animals. Since they no longer had to provide their own armor, ovules then became smaller and lighter than their gymnosperm ancestors, could develop more quickly, and could be fertilized while they were still quite small and soft. This greatly accelerated the reproductive cycle. Gymnosperm pollen tubes have to eat their way slowly through the thick outer tissues of the ovule, a process that can take months.

Animal pollination, simpler stamens, and chambers for the protection of ovules all contributed to the angiosperms' success in environments with short growing seasons, but the ovules themselves underwent the most extraordinary changes. Remember that a haploid, egg-bearing gametophyte must develop within the ovule before fertilization can take place. In gymnosperms, a female gametophyte consists of hundreds of cells and much stored food, in addition to the eggs waiting to be fertilized (Fig. 6.7A). These large ovules take much time and energy to develop, and if they are not fertilized, they drop off, wasting the food stored within them.

In the small soft ovules of angiosperms, however, the gametophyte itself remains in an embryonic state until it is fertilized. It consists of only seven cells (Fig 6.7B) and contains virtually no stored food. One of the cells, located near the ovule tip, serves as the egg, but the oddest part of the story concerns the large central cell, which contains two nuclei.

During fertilization, two sperm cells migrate down each pollen tube, the same as in gymnosperms. One sperm fertilizes the egg, as expected, but the second

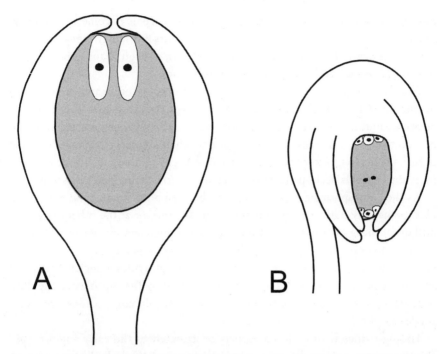

Figure 6.7 Ovules of gymnosperms (A) have one integument and are oriented straight up; those of angiosperms (B) have two integuments and are bent to face downward. The female gametophyte of angiosperms consists of only seven cells; one cell near the ovule opening serves as an egg, and the binucleate central cell will become the endosperm.

sperm is not just a spare in case something happens to the first. It proceeds into the large central cell, where it combines with the two nuclei, creating a triploid cell (one that contains three complete sets of chromosomes), something unheard of in the rest of the living world. This unique process is called double fertilization. The triploid cell then begins dividing rapidly, and nutrients flow into it from the parent plant, forming a specialized food storage tissue called endosperm. Note that the endosperm does not begin to form until fertilization takes place, and in this way, angiosperms don't waste material or energy on ovules that never get fertilized.

The ovules of angiosperms are also peculiarly different from those of gymnosperms in two ways that have no obvious adaptive value. First, each angiosperm ovule (with a few exceptions) is wrapped in not one but two distinct integuments. The two integuments are thin, providing no apparent benefit compared to a single one, so the origin of the second integument is hard to explain. Second, most angiosperm ovules are bent over (anatropous) so that the tip faces back to the base (Fig. 6.7B). A typical gymnosperm ovule stands straight, with its tiny opening (micropyle) at its tip facing straight up (Fig. 6.7A). Though these features seem to have no significance, they may be clues as to who their ancestors were.

Where did angiosperms come from?

When considering where angiosperms came from, we are trying to bridge the gap between other known seed plants, living and extinct, and the earliest angiosperms. What were the predecessors of the unique angiosperm stamens, carpels, and ovules? How and why did these organs evolve? Who were the ancestors of the angiosperms?

Modern phylogenetic studies confirm that angiosperms are not closely related to any modern gymnosperms, though until fairly recently gnetophytes were thought to be their closest cousins (Taylor and Kirchner 1996). Recall that gnetophytes are now considered most closely related to the conifers (Chaw et al. 2000). The split between modern gymnosperms and the line that led to the angiosperms evidently occurred in some ancient group of seed ferns. During the Mesozoic Era (age of dinosaurs), seed ferns diversified and developed many specialized pollen- and ovule-bearing structures. The precursors of the angiosperms must be sought among them.

Several groups of Mesozoic plants derived from ancient seed ferns are fairly well known and were themselves quite diverse: Bennettitales, Caytoniales, Corystospermales, and Glossopteridales. Of course, if the actual precursors of angiosperms lived in dry upland habitats, it is not likely that any were fossilized. The fossils we do have were perhaps descendants of early angiosperms that had migrated into moister environments ("museum" habitats). They can, however, provide clues as to the reproductive features that were developing at that time.

The arrangement of male and female organs into a bisexual flower seems quite similar to structures in the Bennettitales, which flourished from the early Triassic to the late Cretaceous. Members of this order were generally stocky plants with compound leaves, and strongly resembled cycads (Fig. 6.8). The pollen-bearing organs were large compound structures, each with numerous pollen sacs, and these surrounded a central spike of ovules. Some had large bracts below the reproductive organs that could be interpreted as tepals, but in others, the backsides of the large pollen organs fit together to form a hard casing around the whole "flower." This elaborate structure most likely served to channel pollen-eating beetles in an orderly path so as to repeatedly carry pollen from one cone to the ovules in another, just as in cycads and early flowering plants.

The main differences between a Bennettitalean flower and an angiosperm flower are that the ovules were lined up naked along the central axis (there were no carpels), they were straight, and they had only one integument, as in other gymnosperms. The pollen structures could have conceivably simplified into typical angiosperm stamens, but there was no obvious starting point for the peculiar structure of the angiosperm ovules or the formation of carpels. For these reasons,

Figure 6.8 The Bennettitales included plants that superficially resembled the cycads but had rather different and more complex reproductive cones that contained both ovules and pollen-bearing structures. Drawings from Brown 1935.

the "flowers" of the Bennettitales are considered a parallel development and unrelated to those in the angiosperms.

The term "Mesozoic seed ferns" is often applied to a loosely defined group of seed plants descended from seed ferns that shared the earth with the dinosaurs. Like more ancient seed ferns, leaves of these Mesozoic seed plants were complex, and both ovules and pollen sacs were borne on specialized structures that sometimes resembled leaves and sometimes resembled shoots. These extinct orders of plants are considered to be branches of the angiosperm stem group, which in turn probably split off from other seed plants as much as 350 million years ago (Renner 2009). The Bennettitales are sometimes included under the seed fern umbrella, sometimes not. Although the ovules in this specialized group are on simple stalks, the leaves and pollen structures resemble those of earlier seed ferns.

Some other Mesozoic seed ferns did not have flowerlike structures, but in one way or another their stamens, ovules, and ovule-bearing structures are more like those of angiosperms. We do not, however, have enough evidence to clearly point to any particular group as directly ancestral to them. The extinct order Caytoniales has been of greatest interest, but some arguments can be made for ancestry among the Corystospermales and Glossopteridales (Retallack & Dilcher 1981).

In the Caytoniales, pollen was borne on complex branching structures (Fig. 6.9), in sacs at the end of slender twigs, seeming to contradict the notion that early angiosperm stamens were flat and leaflike. In some members of this order, the pollen sacs were bundled in groups of four, suggesting that the angiosperm anther might represent an extreme simplification of such a structure. Stamens would have thus been more like twigs than leaves from the beginning. The pollen sacs in Caytoniales, however, were arranged uniformly around a central axis, not positioned two on each side of a somewhat flattened platform as in angiosperms (see Fig. 6.5B). So we still don't know whether the first angiosperm stamens were twiglike or leaflike.

Ultimately, all reproductive structures in angiosperms and other seed plants trace back to the large fronds of seed ferns. The shoot-like system of branching pollen structures in the Caytoniales probably represents a modification of such an ancient frond. As an analogy, many ferns (such as the cinnamon fern, *Osmunda cinnamomeum*) have specialized spore-bearing fronds that are more shoots than leaves. Finally, the fossil Caytoniales we have are most likely not the direct ancestors of the angiosperms but closely related cousins that may have been specialized for different habitats and different lifestyles. We do not have to try to tie angiosperm structures directly back to any particular fossil form. The various available fossils just represent clues from which we might hypothesize a more generalized ancestor.

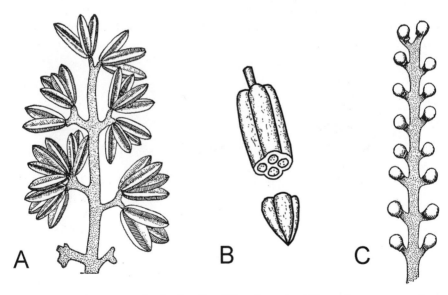

Figure 6.9 In the Caytoniales, both pollen (A) and cupules (C) were borne on complex, branching shoots that were ultimately derived from corresponding leafy structures in ancient seed ferns. In some Caytoniales, pollen sacs are bundled in groups of four (B), reminiscent of the anthers in angiosperms. Drawings from Brown 1935 (A, C) and redrawn after Harris 1937 (B).

The Caytoniales may give us a more direct clue about the origin of the peculiar angiosperm ovules, which with their double integument, simplified gametophytes, double fertilization, and bent configuration, seem to have no precedent. Fossils do not reveal much about the cellular structure or fertilization, but we can see a possible scenario for the double integument and bent configuration.

In the Caytoniales, as well as in the Corystospermales and Glossopteridales, the ovules were enclosed in cupules, which, like carpels, formed protective chambers around the ovules. Upon their discovery, it was proposed that cupules could have evolved directly into carpels. However, the ovules within all of these cupules were like those of other gymnosperms: straight and with just one integument. If cupules had evolved directly into carpels, angiosperm ovules would have remained the same. In addition, cupules were shaped differently: typically more like curved, downward facing vases, with the ovules attached in a single row or cluster along the back side—quite different from the folded structure with two rows of ovules in the angiosperm carpel.

However, the message of the Caytonialean cupule might be completely different. It was curved with a small opening facing downward toward the base (Fig. 6.10A). The ovules were in a cluster attached to the back side of the cupule, and hung downward with their tips directed toward the opening.

This orientation of the cupules led Stebbins (1974) and others to suggest that in some member of the Caytoniales, the cupule became, not a carpel, but the second integument of the ovule. In this particular species, the number of ovules within each cupule was reduced to just one, probably in adaptation to a harsh environment. The small opening of the single ovule then faced downward

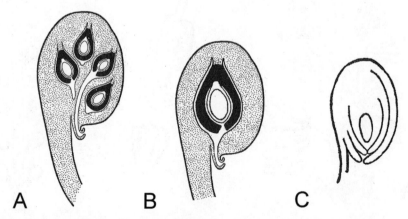

Figure 6.10 According to the Caytonialean hypothesis championed by Stebbins, a cupule with several ovules (A) was reduced to a structure with a single ovule (B), and the cupule wall eventually became a second integument for the ovule (C) through continued reduction. The bent configuration of the angiosperm ovule was thus inherited from the original, downward-facing cupule. Redrawn after Brown 1935.

and more-or-less lined up with the opening of the cupule. Over time, the cupule "shrank down" tightly around the single ovule, effectively becoming another integument (Fig. 6.10B, C). Since the cupule was already bent downward, the new ovule/cupule fusion was also bent downward, just as the typical angiosperm ovule is today. The Caytonialean model simultaneously explains the two anomalies of the angiosperm ovule: its bent orientation and its two integuments.

Recent reviews of all available information (Doyle 2006 and Frohlich & Chase 2007), including new cladistic analyses combining DNA data and morphological characters, support the Caytoniales as the sister group (nearest cousin) of the angiosperms. Though this is still controversial, it considerably strengthens our hypothesis as to how the peculiar structure came about, as well as strengthening the connection with the pollen-bearing structures of the Caytoniales. What about the carpel?

The carpel problem

The classic model of carpel origin is that it evolved from a folded seed-leaf (megasporophyll) with a row of ovules running along each margin. (see Fig. 6.6A). At first, carpels were sealed only by intermingling hairs or by fluids, and pollen tubes may have found access anywhere along the junction. Later, carpels evolved in which the cells along the margins actually grew together, interlocking to form a tight seal referred to as a suture, and only the stigma at the tip provided access for pollen.

Despite giving us invaluable insight on the evolution of the peculiar angiosperm ovules, known fossils from the Caytoniales provide nothing resembling a carpel. As in the pollen-bearing structures, the cupules of *Caytonia* were borne on what appear to be open, branching shoots (see Fig. 6.9C). These shoots were somewhat flattened, however, and might represent modified fronds. Perhaps in a more generalized Caytonealean ancestor megasporophylls were more leaflike, and in that form gave rise to the first carpel. Alternatively, an adjacent leaf might have fused with an ovule-bearing shoot and folded around it.

Amborella and the ANITA grade

In the absence of fossils, we often turn to living archaic organisms for clues. "Archaic" organisms are those whose ancestors split off early from the base of their family tree, and who retain some conspicuous ancient (primitive) features. They are probably not entirely primitive, however. The very fact of their survival for millions of years demonstrates a remarkable success and likely some evolutionary changes along the way. Archaic plants therefore typically exhibit a mix of ancestral and specialized characteristics, and it is sometimes difficult to know which is which. So what can we learn from these relicts of early angiosperm evolution?

Of great interest is the species *Amborella trichopoda*, a shrubby plant inhabiting tropical forests in New Caledonia. Many recent DNA-based phylogenetic studies have confirmed that *Amborella* is in fact the lone survivor of the most ancient clade of flowering plants (see Drew et al. 2014). Its ancestors branched off from the base of the angiosperm tree before any other still-living groups. The next two clades, which branched off soon after, are the water lily order, Nymphaeales, and a relatively obscure group of woody trees, vines and shrubs, the Austrobaileyales (Fig. 6.11). These three groups together constitute the "ANITA grade" of archaic angiosperms.[1] The Austrobaileyales are, like *Amborella*, trees and shrubs with simple flowers, and the two clades together probably represent the general nature of the common ancestor. The presence of the Nymphaeales in this ancient cluster is surprising as its members are all highly specialized aquatic herbs. They are a progressive group that branched off early from the ancestral stock, taking advantage of an opportunity for aquatic life that had previously been the domain of green algae, bryophytes, seedless ferns, and lycophytes. We'll see more about the aquatic lifestyle in Chapter 8.

Amborella clearly possesses some ancient characteristics. Its flower parts, including the carpels, are simple, variable in number, and remain separate from each other. In addition, its wood is composed only of narrow tracheids and parenchyma rays—no vessels or other specialized cells. It may be the only living angiosperm to have inherited such wood directly from ancestral gymnosperms. Other angiosperms, in the Winteraceae for example, also have simple wood without vessels, but this may be a secondary adaptation to cold winters (Feild et al. 2002). On the other hand the flowers of *Amborella* appear to be somewhat specialized, in that they are small and unisexual, and in that the males, at least, are crowded in a mass display. So they might well be specialized in other ways.

In that regard, we are surprised to find that the carpel of *Amborella*, below the stigma at least, is more vase- or urn-shaped than folded. It is a seamless container that appears as if pulled up around its single ovule like a sock (Fig. 6.12). Such a carpel is referred to as ascidiate. There is no suture to suggest the joining of carpel edges, and the one ovule is attached near the top of the carpel—not much like a pea pod with its two long rows of ovules. The carpels of *Amborella* are also unsealed—open at the top just below the stigma, with only a drop of fluid blocking unauthorized entry of dirt, tiny organisms, and dry air.

The implication of *Amborella* is that the original angiosperm carpels were ascidiate and unsealed, and had a single ovule opposite the backbone. It would follow then that folded, leaflike (plicate) carpels with a row of ovules along each margin evolved later from this original design. This scenario is now generally accepted, as stated in the recent paper by Wang and Han (2012). If, however, the

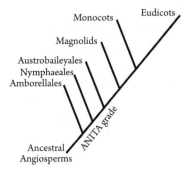

Figure 6.11 In this phylogenetic tree of the major groups of flowering plants, the first three branches at the base are known as the "ANITA grade." The *Amborella* clade consists of a single species, while the three groups at the top consist of numerous families and thousands of species.

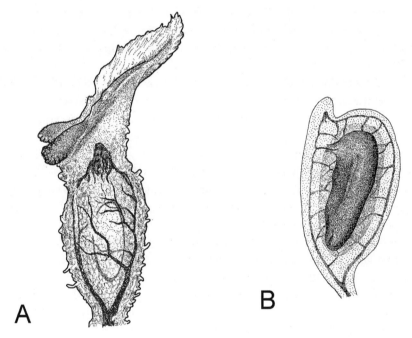

Figure 6.12 In *Amborella*, carpels are bilaterally symmetrical, with a partially folded stigma (A). They do not form a suture opposite the backbone, and so they are considered ascidiate. The mature fruit (B) is a drupe with a single seed. Drawings from Bailey and Swamy 1948.

original carpels were not leaflike with two marginal rows of ovules, then there is a much bigger mystery as to where they came from and how and why they became more leaflike later.

Whatever became the first carpel was most likely flat: either a seed-leaf like those in ancient seed ferns, a leaf below a branching system of ovules, or according to one novel theory, a modified stamen (Frohlich and Parker 2000). The question then is did that blade fold around the ovules, as traditionally believed, or did it grow up and around the ovules in an asciidiate fashion from the beginning?

First of all, there is reason to doubt that the first carpels contained only one ovule. Though that is the case in *Amborella*, other members of the ANITA grade, such as *Austrobaileya*, contain multiple ovules. Moreover, the ovules in the *Austrobaileya* carpel are lined up in two rows, on the side opposite the "backbone," which is what we would expect if the original carpel formed around a flattened, branched structure with ovules on two sides (as in Fig. 6.9C). It could be that the single ovule of *Amborella* is the result of reduction from a more general condition that has survived in *Austrobaileya*. Such reductions have occurred many times, including the reduction to a single ovule in the cupules of some Mesozoic seed ferns (as in Fig. 6.10), and later in the evolution of single-seeded drupes like cherries, plums, and so many others. Drupes are a specialized fruit type with a single, large, food-filled seed adapted for germination in a shady environment, and this is the type of fruit we find in *Amborella*.

Examples of single-seeded drupes evolving into multi-seeded berries are likely to be few, if any, however. Once a carpel has become specialized to service a single large seed, the meristematic activity that could produce a linear series of ovules has typically been suppressed or genetically dismantled. If natural selection favors the production of more seeds in such plants, it is much simpler to increase the number of carpels in a flower, or increase the number of flowers on the plant, by extending the activity of meristems that are already present. This simpler path is predicted by the principle of "adaptive modification along the lines of least resistance," promoted by Stebbins (1974).

There is little doubt, on the other hand, that early carpels were in fact unsealed and that genuine sutures evolved much later (Endress & Igersheim 2000 and Endress & Doyle 2009). In sutures, the edges of the carpel interlock like a zipper. Sutures are absent among the ANITA grade (except possibly in *Illicium*; Endress & Ingersheim 2000, Robertson & Tucker 1979), appearing only in the more advanced magnolids, eudicots, and monocots.

So how do we fit all these clues together? Current thought favors a scenario in which folded carpels evolved from carpels that were originally ascidiate. Something like *Austrobaileya* might have been an intermediate in that transition. Though ascidiate, a new split may have evolved between its two rows of ovules, creating the appearance of folding. Another possibility is that ovules might have been pushed up into the stigmatic zone by a suppression of the ascidiate basal growth. The stigmatic zone then expanded and a suture developed where its two edges came together. This appears to have happened in the unique ANITA grade genus *Illicium*, in which each folded carpel contains a single seed that is dispersed "ballistically" by the sudden splitting of the suture.

If the common ancestor of the crown group angiosperms had carpels with two longitudinal rows of ovules as in *Austrobaileya*, a simpler alternative scenario is possible. The original carpel evolved in the angiosperm stem group as traditionally proposed: through the folding of some leaflike structure, but without sealing by a suture. Most likely as it matured, such a carpel would have simply reopened to release the ripe seeds. Quite possibly, those first angiosperm seeds were equipped with fleshy arils to attract fruit-eating birds, as do the arils that surround the seeds of conifers such as *Taxus, Podocarpus, Cephalotaxus*, and some species of *Juniperus*.

A simple genetic modification could have transferred the fruity function from the aril to the carpel, turning the latter into a brightly colored berry. Like several pivotal evolutionary events described earlier, this might have been achieved by turning on a set of genes in a new location. In this case, the genes for aril development came to be activated during the maturation of the carpel wall instead of during development of the seed coat. This was likely an adaptation for more efficient seed dispersal. Also, since carpels that mature as berries or drupes do not need to reopen to release their seeds, they can be modified to create a more solid and continuous wall and thereby provide better protection for their developing ovules. The early angiosperms did not yet have the ability to form a tight suture at the carpel margins, and so it appears that they achieved this goal by becoming asciidiate, that is by extending growth of the basal tissues of the carpel upwards around the ovules ("as the result of a meristematic cross-zone between the primordium margins"; Endress & Doyle 2009, p. 40), leaving only a small unsealed area in the region of the stigma. The stigma of the carpel in *Amborella* is folded, with flaps on either side of the strong backbone (see Fig. 6.12), which can can be interpreted as a remnant of the folded structure of earlier carpels.

Berries and drupes that evolved later, from carpels with sutures, were already tightly sealed, and so did not become asciidiate. In cherries and plums[2] of the eudicot rose family sutures are still evident along one side of the fruit. So it is possible at least that the asciidiate structure was not a primeval condition in angiosperms but rather an adaptation in the crown group ancestors for closing up the carpel as fruits like berries and drupes evolved.

Asciidiate carpels survived among the ANITA grade, while ancient forms that reopened along carpel margins did not. Perhaps, however, one member of the ANITA grade did retain the ancestral folded structure and persisted long enough to evolve sutures. However they evolved, folded carpels with sutures were so successful that they gave rise to an explosion of magnolids, eudicots, and monocots, which dominate the world today.

Figure 7.1 A visualization by Thomas William Wood of Darwin's hypothesis that a moth with a very long proboscis was the pollinator of *Angraecum sesquipedale* from Madagascar. From Wallace 1867.

7

Adaptations for Pollination and Seed Dispersal

Charles Darwin believed that every feature of a flower had a purpose related to how it was pollinated. We might today use the phrase "adaptive value" rather than purpose, because every plant feature, tiny or outlandish, is hypothetically an adaptation—a change that arose through natural selection in the face of some environmental challenge or opportunity. It is not always easy to discover the adaptive meaning of biological features, and the bewildering variety of flowers provides a huge challenge for that fundamental hypothesis. As Darwin realized, however, the key to understanding flower structure was to understand how flowers are pollinated.

Darwin had a particular interest in orchids. He studied native English orchids for many years and showed that their pollination occurred through intricate interactions with bees and other insects. He also had the opportunity to observe many exotic orchids that had been imported to England. Though he couldn't know their natural pollinators, he could make predictions based on his experience. One orchid from Madagascar, *Angraecum sesquipedale*, was a real puzzler. This fragrant white orchid has an elongate, narrow extension, called a nectar spur, growing out of its lowermost petal (Fig. 7.1). Nectar spurs are common among flowers, serving as a repository for nectar that can only be accessed by animals with specially modified mouth parts, but at 30 cm, this was a whopper.

Darwin knew that other flowers with narrow nectar spurs were pollinated by moths or butterflies (order Lepidoptera), which have mouth parts modified into a slender, liquid-sucking siphon called a proboscis (Fig. 7.2). When moths and butterflies probe for nectar, pollen becomes stuck to their heads or other body parts, and when they visit another flower, the pollen can be transferred to the stigma of that flower. Moths in particular can have rather long probosces, but scientists of Darwin's time knew of none with one 30 cm long, and it seemed preposterous that one would exist. Despite much skepticism among his colleagues, Darwin

Figure 7.2 Moths and butterflies (order Lepidoptera) are among the most specialized but also most numerous of pollinating animals. Their mandibles are highly modified and fused together into a long, narrow tube adapted for siphoning nectar from flowers with narrow tubes, such as this jasmine (*Jasminum* spp.). Drawing from Gray 1879.

predicted that a moth would be found in Madagascar with a proboscis of the right length to pollinate the orchid (Darwin 1862). Eventually it was found—unfortunately years after Darwin's death—in the form of the hawk moth *Xanthopan morganii praedicta*.

The highly specialized and exclusive relationship between *Angraecum sesquipedale* and its hawk moth pollinator is one of many in the orchid family, but there are others even more bizarre, as we'll see later. All are a far cry from the ancient dependence on simple beetles in cycads and early flowering plants, and between these two extremes are many other interesting stories. We can only hit some of the highlights here, but fortunately, there are a number of excellent books on the subject. One of my favorites is that of Meeuse (1961). Though out of date in some ways, it remains a lively, entertaining account of the many relationships between flowers and animals. Other good references on this vast subject include Barth (1991), Faegri & van der Pijl (1979), and van der Pijl and Dodson (1966). For extraordinary visual illustration of many types of flower pollination, Attenboroughs beautiful films in the "Private Life of Plants" (1995, the Birds and Bees episode) can't be beat.

The great pollen giveaway

As you saw in the last chapter, the archaic flowers of *Magnolia* and water lilies are large and wide open to just about any creature that cares to wander in, but it is mostly pollen-eating beetles, flies, and bees that do so, for the flowers provide an abundant supply of nutritious pollen. Colors are subdued in *Magnolia*, but the flowers emit a distinctive fragrance, and sometimes heat, to lure their pollinators (Dieringer et al. 1999). The insects so lured have no interest in the reproductive needs of the flower, and they consume a large percentage of the freely exposed pollen. Primitive beetles added the pollen of early flowers to their older staples from cycads and other Mesozoic seed ferns, and simple flies and bees have since joined the feast.

But what does a plant gain from this expensive giveaway? When the pollen supply in one *Magnolia* flower begins to run out, the insects pull up stakes and fly off to another flower. Inevitably, a few stray pollen grains are stuck to their bodies. Upon arrival at the next flower, the insects land on the highest point, the top of the central column of carpels. Each carpel is tipped by a long, sticky stigma, and as the arriving insects crawl down this elevated staircase toward the mass of stamens at its base, the pollen grains from the previous flower rub off on the stigmas. The insects then begin feeding on the fresh pollen, starting the cycle anew.

So *Magnolia* flowers, as in ancient flowers of similar design, produce excess pollen to attract and reward insect visitors in the gamble that a few grains will make it to another flower. In more recent times, flowers have adopted more sophisticated strategies, like that of the orchids featured in the introduction. Many of these new strategies involve different rewards for the pollinators, commonly a meal of nectar, but sometimes of other substances as well. This diversion saves the more expensive pollen for its more important reproductive function. All flowers, whether rewarding with pollen, nectar, or something else, are precisely constructed so as to take advantage of the habits and movements of their particular pollinators.

Among more modern plants, there are still many that rely upon excess pollen as payment for pollination. Dayflowers (Commelinaceae), peonies (Paeoniaceae), wild roses (Rosaceae), and poppies (Papaveraceae) (Fig. 7.3) all continue in this tradition (Proctor et al. 1996). As in *Magnolia*, such flowers present their pollen in an open, bowl-shaped container, but these more modern flowers add bright colors to their arsenal of attractants. Pollen eaters are typically attracted by bold solid colors, often with pollen of a contrasting color. While beetles remain relatively color challenged, some flies, most bees, and most butterflies have excellent color vision. We will see that color provides a vital role in many of the more specialized types of pollination to follow, often with an expanded toolset of stripes, dots, and other even more intricate patterns.

Figure 7.3 In most members of the poppy family (Papaveraceae), flowers are simple, bowl shaped, and produce large quantities of pollen. Drawing from Kerner & Oliver 1895.

The nectar revolution

A great many angiosperms have switched over to a cheaper and more efficient reward system: nectar. Nectar is basically sugary sap, which runs through plants anyway and so is easy to produce. It can be secreted to the surface at strategic locations within the flower, forming droplets or small pools to be lapped up by appropriately equipped animals. Specialized nectar feeders today include thousands of species of butterflies and moths (order Lepidoptera), as well as many bees, flies, birds, bats, and occasionally other vertebrates, such as lemurs, mice, and lizards.

Nectar production shows up in some members of ancient families, such as pond lilies (*Nuphar*, Nymphaeaceae), and sometimes in families that mostly rely on excess pollen to reward pollinators, such as the Rosaceae (*Prunus, Malus, Potentilla*) and the Papaveraceae (*Fumaria, Dicentra*). Numerous other families rely extensively on nectar for rewarding pollinators, including the mint family (Lamiaceae), the oleander family (Apocynaceae), the several families formerly included in the snapdragon family (Scrophulariaceae), and most members of the orchid and sunflower families. Most of these flowers have petals modified into tubes of one form or another.

The rather broad entry tubes found in most orchids, mints, and snapdragons provide passageways for bees to crawl into the flowers and gather nectar at the back of the tube. Flowers specialized for pollination by butterflies and moths typically form narrower tubes through which only the slender proboscis of the insect can enter. These tubes may be spurs formed from individual petals, as in the *Angraecum* orchid described at the beginning of the chapter (see Fig. 7.1), or from the fusion of the lower parts of the petals into a tube that contains the stamens and the ovary (see Fig. 7.2).

Most moths and butterflies (order Lepidoptera) feed almost exclusively on nectar. Insects in general have a pair of appendages close to the mouth, called mandibles, with which they capture and manipulate their food. In the ancestral moth/butterfly, the mandibles became joined together to form a short proboscis, allowing them to easily suck up liquids. Whether proboscises evolved first for feeding on nectar, for drinking water, or for sucking juices out of fruits cannot be known for sure—all of those behaviors can be seen among moths and butterflies today. But any early flower that began secreting sugary fluids must have attracted ancient fluid-feeding insects and eventually exploited their movements for pollen transport.

Once insects began feeding on nectar, flowers began to form more organized and productive nectaries and also to evolve ways to restrict access to the nectar to those insects most likely to transport pollen to another flower of the same species. Forming slender spurs or floral tubes excluded more generalist insects, saving the reward for those most specifically adapted for the flowers. As floral tubes got longer, moths and butterflies evolved longer proboscises, which in turn led to longer floral tubes. This self-reinforcing cycle of coevolution led to relationships that became more and more exclusive—only particular insects could reach the nectar in the floral tubes of particular flowers. As a result, most moths and butterflies have rather long proboscises, which are coiled and out of the way when not in use (see Fig. 7.2). The relationship between *Angraecum* and its particular hawk moth is an extreme example.

What good is such an exclusive relationship? The hawk moth benefits from the lack of competition—it has a stable, predictable food source. The orchid also benefits from having a single pollinator, because it can be relied upon to move regularly from one flower to another of the same species and thus achieve great efficiency in pollination. The moth's consistent behavior and the precise positioning of the anthers at the top of the floral tube also ensures that the pollen is attached to the same position on the moth's body and will contact the stigma when the moth reaches another flower.

Such exclusivity is common among orchids and, in fact, among most other kinds of flowers, but there is a risk in this strategy. The Hawaiian plant, *Brighamia*, though not an orchid, has a similar exclusive relationship with a native hawk moth (many unrelated plants have adapted to moth pollination). Unfortunately, the particular Hawaiian moth is near extinction, due to human habitat disturbance, and its small surviving populations are not near the surviving populations of *Brighamia*. As a result the plant now has no natural pollinator. It is presently being propagated, as a stopgap measure, by human pollinators dangling from ropes over the cliffside habitat where it grows.

Incidentally, some nectar-feeding flies also have extremely long proboscis. One spectacular example is a tangle-veined fly of the genus *Prosoeca* from South Africa (featured in Attenborough's film). The tangle-veined fly's proboscis is many times longer than the fly's body, but cannot coil like those of moths and butterflies. This makes life difficult, particularly when a breeze is blowing the flowers back and forth. The flies must hover while they maneuver the end of the stiff proboscis into the tiny opening of a flower several inches away. In its South African habitat, there is a succession of long-tubed flowers that sustain the fly through its active season.

Butterfly-pollinated flowers are typically brightly colored and generally lack fragrance. This is because butterflies (and some moths) are active during the day and have excellent color vision. Color is less expensive to produce than fragrance, so it replaced that more ancient attractant in many kinds of flowers as color-sensitive insects evolved and diversified. Butterfly flowers include some with long nectar spurs or floral tubes, but rarely as long or specialized as some of the hawk moth flowers we've seen. Butterflies also feed in small flowers aggregated into dense inflorescences, as in the sunflower family, which have rather short floral tubes. For those nectar resources, butterflies compete with short-tongued bees and nectar-feeding flies.

Many moths are nocturnal. Their flowers are pale in color and highly fragrant. In this way they are superficially similar to archaic flowers like *Magnolia*, but the reason is different—at night time, no one has color vision. Fragrance then becomes necessary. We exploit the fragrances of roses and many other kinds of flowers in the vast scent industry, for traditional perfumes, candles, soaps, and a great variety of other products. Such pleasant scents are vital in particular for nocturnal flowers that are difficult to see, but we'll see later some less pleasant scents at work in other specialized flowers.

The feather-like antennae of nocturnal moths are adapted for detecting even small traces of floral fragrance. The moths can then follow increasing concentrations of the fragrance toward the flowers. The *Angraecum* orchid lacks color but produces a distinctive fragrance attractive primarily to its own specialized hawk moth. The chemical basis of the fragrance is quite complex, consisting of a unique blend of 40 different compounds (Kaiser 1993). Each of the thousands of fragrant orchids, and indeed of all fragrant flowers presumably, has its own proprietary blend recognized by just one or a few pollinators. Nocturnal pollinators therefore are not only highly sensitive to fragrances, but also able to identify the fragrance of their favored food source among the cacophony of smells that waft through the nighttime air.

Speaking of nighttime visitors, the blackest of all known flowers are those of *Lisianthus nigrescens,* a flower from the mountains of central Mexico. Pigments make up 24% of its dry weight, compared with 1.4 % for related purple species (Markham et al. 2004). Why would a plant invest so much in pigments? What kind of animal pollinator would be attracted by a black flower? Perhaps the flower retains heat absorbed by the pigments into the early hours of darkness, and this heat attracts some sort of pollinators, a moth perhaps. But why would this be any better than using white flowers and a fragrance? We still do not know the answers to these questions.

Bees and their flowers

Possibly the most common of all pollinators are bees. Over 16,000 species of bees have been described and named, and most of them feed upon pollen and/or nectar from flowers. Some are generalists, like the common honeybee, which collects both pollen and nectar from a wide variety of flowers, but many bees have developed exclusive relationships with flowers tailored specifically for them.

Flowers adapted for pollination by bees can be just about any color, but are rarely red, because bees don't see red, or not as strongly as other colors. Their typical visual range is from yellow to ultraviolet. Ultraviolet is a distinct color to bees, though it is invisible to us and most other vertebrates. With a specially modified video camera or UV-sensitive film, however, we can "see" UV markings as contrasting dark patches. Some bee flowers may appear to us to be pure yellow, but hidden UV markings help direct the bee into the flower.

Some bees collect only pollen, from large open flowers like most members of the poppy family (Papaveraceae; see Fig. 7.3) and rose family (Rosaceae). Pollen in such flowers is abundant and often brightly colored. Bumblebees, despite their size, are quite nimble around flowers. They can enter a flower, such as a nodding *Clematis*, from the bottom. One has to watch closely, however, to determine whether insect visitors in general, and bumblebees in particular, are truly pollinating the flowers. I have at times observed two species of bumblebee foraging among the flowers of *Clematis reticulata* in Florida, one entering the mouth of each flower to gather pollen, and the other landing outside to pierce the back of the flower where nectar has accumulated. The latter behavior of "robber bees" is fairly common among carpenter bees and bumble bees.

A great many other flowers attract bees only for their nectar. These specialized "bee flowers" most often have petals joined together into a broad tube, large enough for a particular kind of bee to crawl into, with nectar produced at the back of the tube. Most such flowers are oriented horizontally, with the lower part of the corolla extended into a landing platform (Fig. 7.4) marked with conspicuous color patterns to help the arriving bee orient itself. The landing platform may be ruffled or covered with hairs to provide a firm footing for the bees. Flowers modified this way are bilaterally symmetrical (or zygomorphic).

Stamens and stigma are typically located along the top of the tube, and as the bee enters to feed on the nectar, pollen is rubbed onto its back. As the bee moves to another flower, that pollen may be removed by a sticky stigma. The size of the flower and the diameter of the tube are just the right size for the particular species of bee to which it is adapted. As mentioned above, the snapdragon, mint, and orchid families are specialized primarily for this kind of flower. Members of *Iris* and closely related genera are technically radially symmetrical (actinomorphic), if viewed from the top, but consist of three bilaterally symmetrical units jutting out horizontally, each with its own landing platform and corridor leading to the nectar supply.

In orchids, which are mostly bee pollinated, petals are not generally fused together, but the lowest petal (labellum) is enlarged and rolled together at the back to form both landing platform and floral tube. Stamen and style are fused

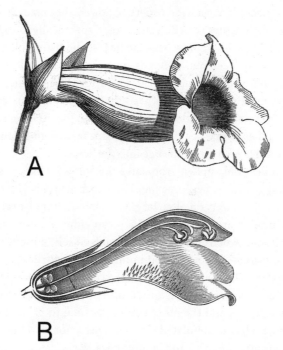

Figure 7.4 The horizontal orientation, broad opening, and marked landing platform of many members of the snapdragon, gesneriad, mint, and other families, are adaptations for pollination by nectar-feeding bees. A. *Ligeria*, from the Gesneriaceae; B. *Leonotus* from the Lamiaceae, showing the stamens and stigma positioned at the top of the flower to contact the back of the bee. Drawings from LeMaout & Decaisne 1876 (A) and Kerner & Oliver 1895 (B).

together and arch over the top of the labellum. The pollen grains in orchids are lumped together into two masses called pollinia, and an insect visitor usually carries away all or none. This adaptation insures full pollination of the exceptionally large number of ovules (over a million) found in each orchid ovary.

Orchids constitute one of the largest and most specialized families of flowering plants. Each species, it seems, has a unique pollinator, and its flowers are uniquely modified to entice, accommodate, and exploit that pollinator. The majority of those pollinators are bees; not your garden-variety honeybees, but wild bees of great variety and habit. While most visit their preferred orchid species to feed on nectar, some feed on nutritious or fragrant oils produced by the flowers specifically for the benefit of their insect visitors. The tropical American orchid *Coryanthes*, for example, secrete a fragrant oil that is gathered by male Euglossine bees and used as a perfume to attract females to them.

Bird pollination

Birds are highly mobile, a primary requirement for pollen transporters, and some have adapted to feeding on nectar. Flowers that are adapted for bird pollination are typically brightly colored, have narrow tubes like those of butterfly flowers, produce abundant nectar, and have no fragrance. Few birds, other than vultures, have much of a sense of smell, and as far as I know, there are no nocturnal nectar-feeding birds. Bird-pollinated flowers are also most often colored red, as birds are highly sensitive to red light, and most insects are not. They can be virtually any color, however, as birds are intelligent enough to recognize flower shape, position, and other cues. Some flowers pollinated by birds are themselves inconspicuous but associated with brightly colored bracts, as in the genus *Heliconia*. This makes pollination more efficient as a series of small flowers can be produced within a single set of more durable bracts.

In the New World, hummingbirds are the primary avian transporters of pollen. Aside from long, narrow, liquid-siphoning bills, hummingbirds have evolved highly specialized wings, with which they can hover motionless in the air. Hummingbird flowers are often angled downward or even dangle completely upside down, making it difficult for other animals to get at the nectar. Familiar examples include the red columbines (*Aquilegia* spp.), red honeysuckle (*Lonicera sempervirens*), and the South American genus *Fuchsia* (Fig. 7.5). The relationship between hummingbirds and their flowers is often quite exclusive.

In the Old World, including Africa and Australia, there are nectar-feeding birds, like African sunbirds, Australian honeyeaters, and Hawaiian honeycreepers, but none have specialized wings capable of hovering in midair. They must perch on the stems of the plants and sometimes have to twist their heads to get into the flowers. But the flowers are typically bright red or orange. Examples

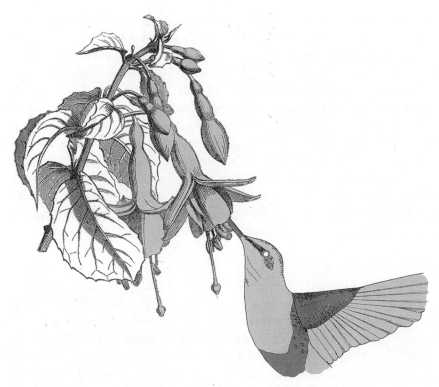

Figure 7.5 The genus *Fuchsia* consists of many species native to the Andes Mountains of South America. Most are colored bright red and/or purple and hang downward. They are pollinated primarily by native hummingbirds. Modified from LeMaout & Decaisne 1876.

include the familiar genus *Aloe,* an occasional African heather (*Erica*), *Protea* and other African genera in the Proteaceae, and *Banksia* and *Grevillea* in the same family from Australia.

Mammals as pollinators

Nectar and pollen are both highly nutritious materials, and a number of vertebrates in addition to birds have taken to feeding in flowers. Amongst mammals, bats are attracted at night to large, fragrant, pale-colored flowers with an abundance of pollen and/or nectar. Many cactus flowers, including those of the giant saguaro (*Carnegia gigantea*), are pollinated by bats (Fig. 7.6). Bats, of course, fly like birds or insects and therefore are highly effective in dispersing pollen.

Nonflying animals are rarely pollinators, but a bush mouse in South Africa and lemurs in Madagascar have specialized pollination relationships with certain plants. The bush mouse laps up nectar from low-hanging flower clusters of a species of *Protea,* while lemurs in Madagascar travel great distances through forest

Figure 7.6 Many night-blooming cactus flowers, as in the saguaro, are pollinated by bats who come to feed on nectar and pollen.

canopies carrying pollen from one traveler's palm (actually a bird-of-paradise relative, *Ravenala madagascariensis*) to another. Both of these are nicely illustrated in Attenborough's films.

Flowers that deceive

There is a class of rather gruesome pollination syndromes that involve flowers that mimic corpses of animals—the carrion-flower syndrome. This is a form of deception in that the insect visitors receive no reward at all. In carrion flowers (or inflorescences in some cases) the petals or bracts are a mottled brown in coloration and emit a foul odor, creating an illusion of rotting animal flesh that attracts carrion flies and other insects that feed and often breed in actual corpses.

These flowers offer no reward, but carrion flies and beetles are apparently slow learners and may visit several such flowers before finding an actual corpse where they can feed and/or breed. In doing so, they carry pollen from one flower to another. Carrion flowers can be found in great variety among the succulent members of the milkweed family (Asclepiadaceae) (Fig. 7.7), some members of the pipevine family (Aristolochiaceae), the inflorescences of many aroids (Araceae), and in the largest flower in the world, *Rafflesia* (Fig. 7.8).

Figure 7.7 *Stapelia* is one of many cactus-like members of the milkweed family (Asclepiadaceae) found in Africa, with foul-smelling flowers variously colored to resemble dead animals. Drawing from LeMaout & Decaisne 1876.

Figure 7.8 Giant *Rafflesia* flowers appear unexpectedly from parasitic underground plants growing in Indonesian rain forests. Drawing from Kerner & Oliver 1895.

An even crueler deception can be found among orchids, which seem to be popping up throughout this chapter. Among their many specialized adaptations for pollination, the most bizarre have to be those that exploit weak-minded male bees and wasps in pursuit of mates. Orchids of the European genus *Ophrys*, for example, have evolved to strongly resemble the coloration, texture, and even scent

Figure 7.9 The wasp orchid, *Ophrys speculum*, mimics both the appearance and the sex pheromone of the wasp, *Campsoscolia ciliata*. Hairs around the margins of the flower resemble those on the abdomen of the wasp, and the middle tepals resemble wings. The large clear area of the lower tepal also has a shimmering, bluish tone evoking the visual cue of the wings.

of female bees and wasps (Fig. 7.9). The lower lips of these orchids are frequently fringed with reddish-brown hairs and may have bluish patches that resemble insect wings reflecting sunlight.

The deception is good enough to draw male bees from miles around. Males of the particular bee or wasp species are drawn to these flowers expecting to copulate with females. As they wiggle and writhe around on the phony mates, packets of pollen (pollinia) become stuck to their heads (or sometimes to their rear ends), and these can be deposited on the stigma of another orchid flower when the befuddled males are tricked into another episode of "pseudocopulation" (see Proctor et al. 1996; also Attenborough 1995). Orchids in other parts of the world, particularly Australia, have adopted similar deceptive practices to get their pollen transferred.

Plants with many small flowers

The pollination adaptations of moderately large, conspicuous flowers are relatively easy to decipher. Many angiosperms, however, have small flowers that are individually inconspicuous, but when massed together, they form a showy display. How are they pollinated? Careful observation is required to find the answer.

In the jungle swamps of Costa Rica grow several species of the palm genus, *Bactris*. These slender palms are covered with spines and have tasty purple fruits. *Bactris* flowers are small and lined up by the hundreds or thousands on the branches of specialized inflorescences. Each flower by itself would scarcely warrant the attention of any passing insect, but together they create a display far more massive than the large flowers of *Magnolia* or any of the bat-pollinated flowers. In *Bactris*, moreover, the flowers are unisexual, and their development is highly synchronized. On a particular inflorescence, all the flowers mature at the same time, but male and female flowers open on different days. Such highly integrated inflorescences function like superflowers, with the male and female flowers functioning more or less like the individual stamens and carpels in a simple flower.

The action begins each day as it starts to get dark. Within a population of these palms, there are some inflorescences that have been open all day, and some that are just beginning to emerge from the thick bracts that enclose them during development. This new crop of inflorescences begin opening around 4:30 in the afternoon, and are fully expanded by 6:30 or 7:00 p.m. At that time, all the tiny female flowers, which consist of little more than a pistil and a few bract-like petals and sepals, are open. The stigmas are expanded, moist and glistening, ready to receive pollen. The male flowers attached on either side of the females have large, fleshy petals that are still tightly closed over the stamens. Soon, large numbers of insects begin to arrive, mostly small flies and beetles called weevils. The weevils come and begin to feed on the succulent male flowers, showing no particular interest in the female flowers. Many of them have pollen stuck to their bodies from the previous inflorescence visited. and as they crawl around the inflorescence, they sometimes pass over the stigma of a female flower and deposit a grain or two of pollen.

These insects have come from nearby inflorescences that had opened the evening before, and in which the male flowers have just opened to release their pollen. On those day-old inflorescences, the stigmas of the female flowers are already withered. The weevils have spent the last 24 hours feeding on the unopened male flowers, and they begin leaving in droves when the flowers open. As the weevils migrate to the newly opened inflorescences, they inevitably carry a few pollen grains with them. This cycle repeats itself night after night through the blooming season (Essig 1971).

Pollination of an entire inflorescence thus occurs quickly and efficiently, and by separating the activity of the male and female flowers by 24 hours, self-pollination is avoided. Very similar observations were made on a species of *Hydriastele* in Papua New Guinea (Fig. 7.10) (Essig 1973). Temporal (timing) separation of male and female organs is common among a variety of flowers, though other mechanisms for avoiding self-pollination have evolved. Some flowers have a type of self-recognition, in which pollen grains that land on the stigmas of their home plant fail to germinate or develop properly.

Ironically, pollination of the large but simple *Magnolia* flowers and the complex inflorescences of *Bactris* is accomplished by similar small insects, despite the striking difference in the size of the flowers. The dense spikes of flowers in these and many palms, in fact, harken back to the cones of cycads, which are also

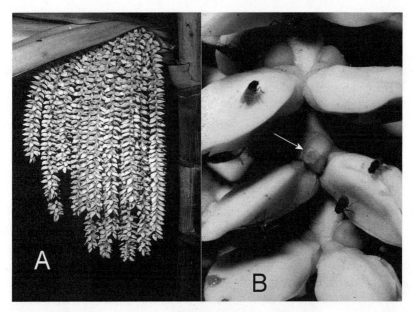

Figure 7.10 Like *Bactris* in Central America, the flowers of *Hydriastele*, found in Papua New Guinea, are small and numerous (A), and typically pollinated by pollen-eating flies and beetles (B). The small female flowers (arrow) are all receptive as soon as the inflorescence opens, while the larger male flowers on either side remain closed until the next day.

pollinated by similar insects. It would seem that evolution has come full circle. Are superflowers better than large single flowers?

One thing we can say is that palms are more reproductively prolific than *Magnolia* trees. A palm inflorescence produces hundreds of female flowers, and in inflorescences that behave synchronously as *Bactris* inflorescences do, the rate of fertilization is 90% or better. A *Magnolia* flower, of which there may be several dozen on a tree at a time, may contain several dozen carpels, and their fertilization rate is typically 40–50% (Dieringer et al. 1999).

The similarities between *Magnolia* and *Bactris* pollination strategies are clearly due to convergence, but similar syndromes have evolved many times among the flowering plants. Flowers in many families have been reduced and aggregated into massive inflorescences, which are visited by small beetles and other simple insects like flies and bees. Striking examples are found in the Araceae (calla lily family), Apiaceae (carrot family, including dill and celery), Cyclanthaceae (Panama-hat palm family), Asteraceae (sunflower family), Oleaceae (olive family, including privet), and Piperaceae (piper family, including *Peperomia*—Fig. 7.11A). Where only pollen and succulent flower parts are available, simple beetles, bees, and flies predominate as pollinators. If nectar is also produced, as in the Apiaceae, Asteraceae, and Oleaceae, nectar-feeding flies, bees, moths, and butterflies will also be present.

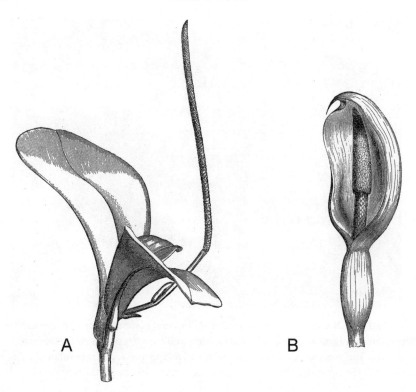

Figure 7.11 The flowers of *Peperomia* (A) are tiny and crowded on a long spike. The flowers of an aroid (family Araceae) (B) are also tiny and densely packed on a central spike that is surrounded by a conspicuous bract, resulting in a whole that functions as a compound superflower. Drawing from LeMaout & Decaisne 1876 (A) and Kerner & Oliver 1895 (B).

Compound flowers

Taken a few steps further, masses of tiny flowers can become what we call "false flowers" or compound flowers. In a compound flower, a head of small flowers is surrounded by bracts or other structures modified to look like petals. The whole compound structure then looks like a single large flower.

Inflorescences of the aroid family (Araceae, including calla lilies, philodendrons, anthuriums, etc.) can be considered compound flowers. The central axis of the aroid inflorescence (the spadix) is packed with tiny flowers—even smaller than those of palms—that also function synchronously to act as a single superflower (Fig. 7.11B). Some aroids have spikes of bisexual flowers, while the more specialized have female flowers confined to the lower part of the spadix with male flowers above. A large bract, or spathe, around the spadix protects the flowers during development, and it is often colored in such a way as to attract pollinators. The whole looks like a large and rather bizarre flower.

The largest aroid inflorescence in the world is that of the titan arum (*Amorphophallus titanum*) of Indonesia. The spadix may be 3 meters high and the spathe around it may spread to nearly 2 meters. It attracts small sweat bees that spend the night in the lower chamber where the female flowers reside. During the night, male flowers above them shed their pollen, and as the bees leave in the morning they become covered with pollen and carry it off to another inflorescence.

Other families that include tightly integrated inflorescences that resemble single flowers include the Proteaceae, which is native primarily to South Africa and Australia, the Cornaceae (dogwoods), the Euphorbiaceae (poinsettia family), and most important of all, the Asteraceae (sunflower family). The sunflower family is arguably the most successful family of plants currently on the planet, if you measure by the number of species. With 1620 genera and 23,600 species (Stevens 2001 onward), it is one of the largest families of plants (the orchid family may be larger with estimates ranging from 22,000–35,000 species). It is certainly one of the most ecologically diverse, being particularly abundant in seasonally dry or cold climates, and often dominating open spaces with a succession of species throughout the growing season.

Sunflower heads typically (but not always) consist of two kinds of flowers (Fig. 7.12A). Tiny disk flowers occupy the center of the head, and these are the sexually functional flowers. Around the edge of the disk is a series of ray flowers (Fig. 12B), which are often sterile, but which have a very long, flat extension of the corolla

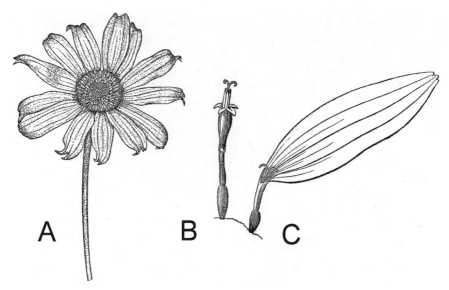

Figure 7.12 A. In the sunflower family (Asteraceae), flowers are reduced and crowded into a dense head; B. flowers in the center (disk flowers) are typically radially symmetrical with short tubular corollas; C. those around the edge (ray flowers) typically have a corolla modified to resemble the petal of a large flower. Drawings from Brown 1935 (A) and Gray 1879 (B, C).

that looks like a single petal. Beneath them, one or more series of green sepal-like bracts, called phyllaries, complete the illusion of single, large flowers. The entire structure looks like a flower with many petals, and attracts a variety of unspecialized pollinators that feed on pollen and nectar. Some members of the family have heads consisting of only ray flowers, and some consist only of disk flowers, but in most cases the flower heads are easily mistaken for single flowers.

Brood-space pollination

Some flowers and densely packed inflorescences provide not only food but also breeding accommodations for insects. Figs are the most famous example. The fig genus (*Ficus*) consists of over 600 wild species, some of which are known as banyans or strangler figs, as well as the common cultivated fig (*Ficus carica*). Fig trees are dependent on pollination by tiny wasps that breed inside the figs. The fig is an "inside-out" inflorescence. During its evolution the inflorescence inverted so that the flowers came to reside inside a small round chamber with a tiny opening at the tip (Fig. 7.13). Female wasps, with eggs already fertilized enter young figs and lay their eggs inside. The larvae develop, nourished by the tissues of the fig, and then hatch. Male wasps hatch first. They are wingless and have only one task to accomplish in a very short life. They seek out the female larvae still embedded in their chambers, copulate with them, and then die.

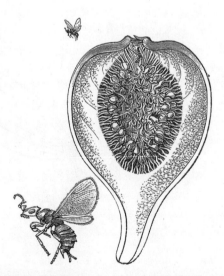

Figure 7.13 A fig is a highly specialized chamber containing many small flowers. The wasp *Blastophaga grossorum* lays its eggs within the chamber. The next generation of young wasps leave the chamber just as the male flowers mature and carry pollen to another fig. Drawing from Kerner & Oliver 1895.

The female wasps continue their development for several days, eggs already fertilized within them, then emerge. They have wings and seek out the small opening of the chamber. At about the time they emerge, stamens of the male flowers split open to release their pollen, and as the females leave they are covered with it. When they enter another fig to begin the cycle again, they pollinate the female flowers within. Incidentally, the cultivated fig produces fruit without pollination, so if you were about to swear off of them for fear of getting a mouthful of wasps, rest assured.

Figs are not the only plants that provide brood space for insects, though this pollination syndrome is relatively rare outside of that family. In the mangrove palm, *Nypa fruticans*, of southeast Asia and Pacific islands, tiny fruit flies lay their eggs in the dense, fleshy spikes of male flowers. As the young flies emerge a few days later, they are covered with pollen and fly off to find another inflorescence. When they do, they land on a head of female flowers, which emerge first, and deposit some of their pollen load on the stigmas. The strategy is remarkably similar to that employed by cycads. Incidentally, this is another bit of information for which the world owes my standing for hours out in a swamp—this time in Papua New Guinea, but fortunately not at night (Essig 1973).

The famous yucca moth (genus *Tegeticula*) lays its eggs in the ovaries of the flowers of the desert *Yucca*, where the young complete their development. *Yucca* flowers are ordinary looking, not reduced and contained in a chamber like those of figs, but are dependent on the yucca moth for pollination. In fact, this may be the only known case of "deliberate" pollination on the part of the animal. In most other pollination syndromes, the transfer of pollen is accidental as far as the animal is concerned, but the yucca moth is programmed to actively insert a lump of pollen between the closely pressed stigmas of the *Yucca* flower, after first having checked to determine whether eggs had already been laid there. After it lays its eggs, the moth gathers up pollen and repeats the process in another flower. How does this benefit the *Yucca* if its ovary is consumed by a moth larva? The moth is programmed to deposit eggs in only one of the three ovary chambers of the *Yucca* flower, and the other two are left to develop seeds normally (Ingrouille & Eddy 2006, p. 169).

Trap flowers

There are a number of pollination syndromes that involve "traps." We've seen some already in water lilies. In each of these, insects are temporarily trapped within a flower or within specialized inflorescences. Sometimes they are forced to exit through elaborate passageways where anthers and stigmas are located, or to exit the flower the same way they came in after some time has passed and the flower has switched from female to male mode. Variations on the trapping strategy occur in the Dutchman's pipe (*Aristolochia*), a number of orchids, and sweetshrub (*Calycanthus*).

"Trap inflorescences" occur in many members of the Araceae, such as the jack-in-the-pulpit (*Arisaema*) and the dead horse arum (*Helicodiceros muscivorus*), which also employ the carrion deception we saw earlier. Both emit fetid odors that attract carrion or "mushroom" flies, but the dead horse arum is most blatantly a bogus corpse. Both have a restriction between the upper male and lower female flowers (see Fig. 7.11B). Insects enter during the female phase, wandering down the spadix following a food-promising fragrance. They cannot exit because of a barrier created by the constricted spathe and/or a ring of spikes that prevent them from crawling upward. They spend the night, and in the morning, the spikes have withered and the inflorescence has entered the male phase. As the insects crawl out they are covered with pollen and fly off to another inflorescence in the female phase.

Wind pollination

Bisexual flowers evolved in the earliest angiosperms and served as a more efficient way to have pollen carried from one plant to another by animals. Nevertheless, a great many flowering plants have turned to wind pollination, for in some habitats, it is still the most efficient means of pollen transfer. Many trees of the broadleaf deciduous forest of the northeastern United States and Eurasia employ wind pollination in the early spring when few insects are available and while leaves are still absent. Oaks, elms, beeches, birches, walnuts, hickories, some maples, and a host of others are wind pollinated (Fig. 7.14). In the monocots, herbaceous plants of open meadows and plains, most particularly the grasses and sedges, are wind pollinated, as are some semiaquatics, like rushes and cattails. These plants inhabit dense stands that may stretch for hundreds of miles, and so wind pollination works quite efficiently.

Flowers that are wind pollinated mostly produce only one seed per flower, because the odds of a single flower catching more than one pollen grain from the wind are small. They also lack the various trappings required for attracting animal pollinators: color, fragrance, and nectar. Petals are reduced or absent, but stamens are large and flexible and produce large quantities of light pollen. Stigmas are large and often feathery or comblike for filtering out pollen grains from the wind (Fig. 7.15).

Wind-pollinated flowers also tend to be unisexual, with male and female flowers separated on different parts of the tree or on different trees, as an adaptation for avoiding self-pollination. In other words, wind-pollinated angiosperms have become more like conifers. Some even have their flowers in cones (as in *Betula*, *Casuarina*, and *Alnus*), a fact that led some early evolutionary taxonomists to believe that the first angiosperms were descended from conifers. Many grass

Figure 7.14 Many trees of the temperate deciduous forests, such as this hazelnut (*Corylus*), are pollinated by the wind. The large female flowers (left) are in small clusters, while the tiny male flowers (right) are in long, flexible catkins. Drawing from Kerner & Oliver 1895.

flowers remain bisexual and therefore potentially self-pollinating, though probably avoiding that by ripening anthers and stigmas at different times. Some species are self-fertile, a fact that allows them to rapidly colonize disturbed sites.

Water pollination

Many flowering plants have reverted to a lifestyle even more ancient than that of wind-pollinated gymnosperms. They have become aquatic and in some ways like algae. While some hold their flowers above water for ordinary forms of animal or

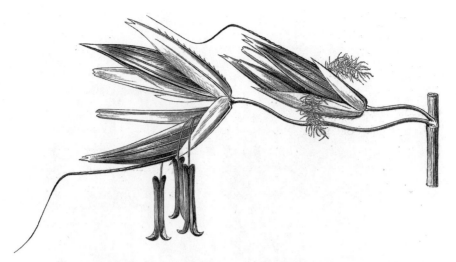

Figure 7.15 Grass flowers have reduced petals and sepals and are surrounded by green or inconspicuously colored bracts. Stamens dangle below the flowers, where the wind can readily pick up the dustlike pollen. Stigmas are large and feathery, serving as combs to filter pollen out of the air. Drawing from Kerner & Oliver 1895.

wind pollination, others are pollinated underwater or at the surface of the water. In all, some 31 genera in 11 plant families exhibit water pollination (Ingrouille & Eddy 2006). Sea grasses, occurring in some 12 genera in 4 angiosperm families (e.g., *Thalassia, Zostera*) produce pollen in long threads or slimy mats that drift in the currents until they catch on the stigma of another plant. Sea grasses, as their name suggests, mirror the lifestyle of terrestrial grasses, substituting water currents for air currents in pollination. Sea grasses are not closely related to terrestrial grasses, however. Both are monocots, but they evolved from different lineages.

Elodea, a common aquarium plant, has flowers on long stalks that reach to the surface. Male flowers release pollen grains that float on the surface until they contact the stigma of a female flower. Flowers of *Vallisneria* go one step further. These plants are rooted in the mud below the water, and the female flowers rise to the surface on long stalks. The male flowers, however, break off from the underwater rhizome and rise by means of gas-filled blisters to the surface, where they float freely. Male flowers that drift close to the much larger female flowers are drawn inside by surface tension, and their stamens contact the stigmas (Fig. 7.16).

Ripe fruits participate in the dispersal of seeds

Flowering plants, like other seed plants, have two opportunities within their life cycle for dispersal of their genes. This is another advantage of seed plants over those that disperse only by spores. The first opportunity is when the pollen grains

Figure 7.16 In the aquatic plant, *Vallisneria*, male flowers detach from the underwater rhizome, float to the surface, and by chance contact the larger female flowers that remain attached to long stalks. Drawing from Kerner & Oliver 1895.

are released and carried around by wind or animals, often for many miles. The second is when the ripe seeds are likewise spread long distances by animals, wind, or water. Dispersal is important for colonization of new areas, escaping from dense herbivore populations, and mixing genes among widespread or isolated plant populations.

In the flowering plants, seeds develop within fruits. The fruit is the mature carpel or pistil, which may be radically transformed as the seeds ripen, and usually plays a direct or supportive role in the dispersal of the seeds. Though we tend to think of fruits as fleshy, sweet, and edible, they can also be dry and hard or papery. In Chapter 6, we considered the question of whether the first fruits split open to release their seeds, or instead became fleshy berries. In either case, reptiles or birds were common dispersers of seeds at that time.

Fruits that split open at maturity release the seeds for dispersal on their own. The first angiosperm seeds may have attracted animal dispersers with fleshy arils, but many more mechanisms have evolved. A form of deception occurs on seeds that appear to have an edible aril on them, but which is actually just a colored portion of the hard seed coat. These are referred to as mimetic fruits. They are common in the legume family (Fabaceae) and include the notoriously poisonous rosary bean, *Abrus precatorius*.

Commonly, seeds have tufts of hairs, umbrellas, parachutes, or wing-like extensions (Fig. 7.17), which aid their dispersal by wind. Orchid seeds are small and dust-like and waft away even in the slight breezes of a tropical rain forest. Commercial cotton, and a similar material from the unrelated cottonwood tree (*Populus*), is derived from tufts of hairs originally adapted to keep seeds aloft in the wind.

Figure 7.17 In dry fruits that open at maturity, the seeds are variously adapted for dispersal on their own. In this trumpet vine (*Bignona* sp.), seeds have flattened, wing-like extensions that aid their drifting in the wind. Drawing from Kerner & Oliver 1895.

Seeds in other dry capsules may simply rattle out when swaying to and fro in the wind (as in poppies) or when bumped by a passing animal. Some may stick to the fur or feathers of animals. Other fruits develop tension in their walls as they dry, which builds up until they literally explode and expel their seeds some distance, for example, many legumes, geraniums, members of the mustard family (Brassicaceae), and the sandbox tree (*Hura crepitans*). The common *Impatiens*, and the squirting cucumber, *Ecballium elaterium*, are fleshy fruits but are adapted for forceful ejection of the seeds rather than to be eaten.

Fleshy fruits are adapted to be eaten

Before angiosperm fruits, many gymnosperms had fleshy coverings over their seeds, either the outer layer of the seed coat, as in cycads and ginkgos, or a special fruit-like layer called an aril that develops around the seed, as in the yews (*Taxus*). Early angiosperms thus had readily available fruit dispersers and could have exploited them either with berries or with dry fruits bearing seeds with brightly colored arils.

The seeds of fleshy fruits are consumed along with the fruit tissues, as are seeds with arils, and if relatively small, they pass through the digestive tract and are eventually deposited with the feces. This may be miles from the parent plants, and so the seeds are effectively dispersed. The seeds are resistant to the digestive processes of the animal and may even benefit from the acidic environment, which can help break down a hard seed coat. The small seeds of grapes, blackberries, watermelons, cacti, and many others travel in this way (Fig. 7.18). Birds are the principle dispersers of such seeds, but bats and monkeys are also major fruit/seed dispersers in the tropics, especially of the many species of figs. Reptiles, such as tortoises, disperse the seeds of cactus fruits, but obviously more slowly.

Birds that consume fruits at least some of the time are too numerous to list. They include familiar temperate birds like pigeons, robins, orioles, chickadees, wrens, and cardinals, and also tropical birds like toucans, hornbills, cassowaries, and parrots. As noted when discussing bird pollination of flowers, modern birds are attracted by bright colors, particularly red, and that applies as well to fruits

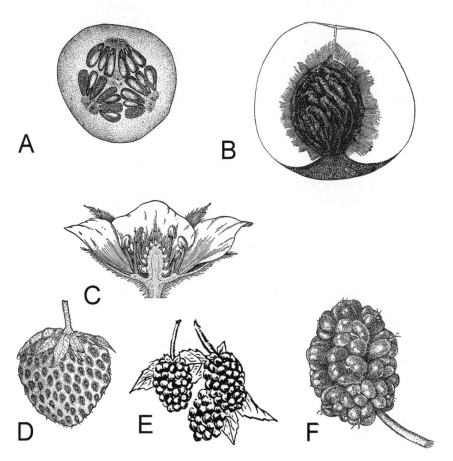

Figure 7.18 Fleshy fruits contain edible tissues, which animals consume: A. Cucumber; B. Peach; C. the flowers of strawberry (*Fragaria*) and blackberry (*Rubus*) are virtually indistinguishable, but the fruits (D, E) develop differently; F. mulberry. Drawings from Brown 1935 (A, C, D, F), LeMaout & Decaisne, 1876 (B), and 4vector.com free clipart (E).

that are adapted for dispersal by birds. The bright red berries of holly and other temperate plants that appear in the fall and winter sustain many species of birds when little else is available to eat. Red berries are also common in the tropics, on palms, fig trees, and many others.

Cassowaries in New Guinea and northern Australia, however, show a preference for the blue fruits of the "Cassowary plum" (*Cerbera floribunda*). This fruit has juices poisonous to most other animals, but the cassowary is immune to the poison as an adaptation to this remarkable symbiosis. The bird is nourished while the seed is dispersed. The large seed is not disturbed as the fruit passes through the cassowary's digestive system and appears to germinate better after such a journey. Seeds of the rhino apple, *Trewia nudiflora*, are dispersed in the same way by Indian white rhinoceroses along floodplains in Nepal.

The cassowary plum and the rhino apple are examples of fruits with a single large seed rather than many small ones. Similar fruits are found in *Prunus*, the familiar genus of cherries, peaches, and plums, as well as in most species of palm. Such seeds are typically enclosed in a hard pit that forms from the inner fruit wall, in which case they are referred to as drupes. The large seed of an avocado, however, is embedded directly in the soft flesh of the fruit, so is not a drupe, but technically a large, single-seeded berry. That seemingly trivial distinction serves to illustrate the fact that very similar structures can evolve in response to similar conditions, but with different details of structure. This indicates convergent evolution from ancestors with different kinds of fruit. The larger seeds of drupes and avocado contain a large amount of stored food, an adaption for germinating in shady forests where the young plants may have to grow several feet high before reaching sufficient light to sustain themselves.

These larger seeds can pass through the digestive tract of large animals, as we've seen in cassowaries and rhinos. In some cases, smaller birds feed on drupes in which the seeds are too large to pass through the intestine, in which case the seeds are regurgitated after the soft fruit tissues have been digested off (Hernández 2009). In Florida, cherry laurel seedlings can be seen sprouting up in great numbers along the bases of chain-link fences, a favorite perching site for birds digesting meals of berries and drupes.

As noted in Chapter 6, many archaic angiosperms of the ANITA grade have brightly colored fleshy berries adapted for consumption by birds. Such fruits are common among magnolids, eudicots, and monocots as well, testifying to the great success of this coevolutionary collaboration with fruit-eating birds. In the rush to take advantage of this seed transportation system, many false fruits have evolved as well. A strawberry is not a fleshy fruit, but rather an expanded receptacle upon which sit dozens of tiny, hard, dry fruits called achenes. Each achene contains a single seed. The strawberry achieves exactly the same goal as a true berry, despite its inside-out structure.

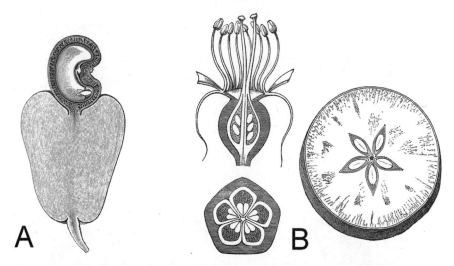

Figure 7.19 False fruits: In the cashew (A), the receptacle below the nut is swollen into a fruit-like body; in apples and pears (genus *Pyrus*; B), the receptacle extends up and around the carpels and swells into edible fleshy tissues. Drawings from Kerner & Oliver 1895.

Other false fruits include figs, and ironically, apples—the quintessential fruit of history. Figs and apples are false fruits in the sense that at least part of the edible material does not come from the carpels. A fig is a whole inverted inflorescence, with many small flowers inside. The fleshy wall of the "fruit" is the highly modified, hollow upper end of the flower stalk (the receptacle), turned inside out like a sock.

The apple is similar, but simpler. It comes from a single flower, but most of the edible fleshy tissue comes from the top of the receptacle (Fig. 7.19) that has grown up around the carpels. The true fruits, the hard chambers containing the seeds and a thin layer of flesh around them, are deep inside.

Dry fruits that do not open

A vast array of fruits are dry, hard, papery, spiny, or otherwise unappetizing, but also do not open to release their seeds. Usually they contain only one seed, and the fruit takes over the dispersal roles otherwise performed by the seeds themselves. Some, for example, are wind dispersed. Maple fruits and many others, for example, have "wings" for wind dispersal. Umbrellalike structures allow small dandelion fruits to drift on breezes, as do the feathery tails of *Clematis* achenes (Fig. 7.20).

Nuts and grains are single-seeded fruits in which the dry, hard, fruit wall, or part of it, permanently adheres to the seed. Acorns (*Quercus*) are good examples,

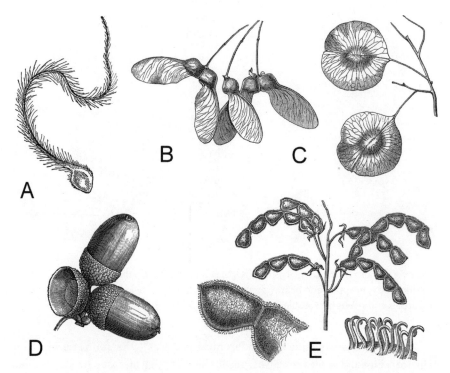

Figure 7.20 Nonfleshy fruits may be adapted for a wide range of dispersal mechanisms: In *Clematis* (A), *Acer* (B), and *Ptelea* (C), extensions of the fruit allow for wind dispersal. Nuts such as the acorn (*Quercus*) (D) are hoarded by animals, and some survive to sprout. In *Hedysarum* (E), tiny hooks on fruit segments allow them to catch on animal fur. Drawings from Ganong 1916 (A) and Kerner & Oliver 1895 (B–E).

as are walnuts, filberts, and hazelnuts. These seeds are large and filled with food reserves. Of course, many animals seek them out and consume them. This would seem to be a serious disadvantage to the plants, but it is actually part of their adaptation for dispersal.

Nuts and grains have no special adaptations for dispersal and are generally heavy enough that you would expect them to simply fall to the base of the tree and not be dispersed at all. Their strategy is similar to that of flowers that produce large quantities of pollen: lots are produced in the chance that a few will complete their mission. Oaks and grasses alike produce so many seeds that the animal populations cannot consume them all. Squirrels, chipmunks, and other rodents, for example, hoard away nuts and grains to sustain themselves through the winter, often by burying them far away from the parent tree. In the spring, any buried nuts that have not yet been eaten sprout and develop into a new generation.

Dry fruits, like fleshy fruits, have evolved many times and in many different ways, resulting in striking instances of convergent evolution. Walnuts, acorns, filberts, and pistachios are all in different families, as are winged maple and ash fruits. This array of seed dispersal mechanisms, along with diverse pollination syndromes, has contributed enormously to the success and diversity of the angiosperms. As we'll see in the next two chapters, an array of vegetative adaptations adds a third dimension of variables, increasing that diversity even more.

Figure 8.1 Dicotyledonous trees like this oak dominate the temperate forests of the northern hemisphere. Drawings from Kerner & Oliver 1895.

8

The Dicotyledonous Grade

When European settlers first arrived in North America, they encountered familiar forests of massive oaks, maples, elms, and other hardwoods (Fig. 8.1). Such forests figure prominently in cultures that span the northern hemisphere from western Europe into Russia, China, Korea, Japan, and across the ocean to North America. The painting, poetry, music, and fiction of these cultures revolve heavily around the changing of the seasons, reflected most conspicuously by the changes in the trees. At the first frosts of fall, the broad leaves of most of these trees turn brilliant hues of red, orange, and yellow as their green chlorophyll molecules disintegrate, and their minerals are recycled. By the onset of winter, the leaves have fallen, leaving the trees bare and dormant.

Settlers from Spain and Portugal arrived in warmer parts of the New World, also encountering vast forests of broad-leaved trees, but here of a bewildering and unfamiliar variety, and mostly evergreen. In drier regions, they found plants with more specialized forms: trees and shrubs with small, hard, drought-resistant leaves; leafless plants with water-filled, photosynthetic stems; or plants with large, water-filled leaves; all adapted to conserve water in the face of intense sunlight and high temperatures.

These dominating plants belong mostly to the largest clade of flowering plants, the eudicots. Formerly, the angiosperms were divided into two clearly defined subclasses: the Dicotyledonae and the Monocotyledonae. Because of changes in the modern rules of taxonomy,[1] the dicotyledons are no longer a formal taxonomic category. For many practical purposes, the eudicots are roughly their equivalent, but the dicots also included the ANITA grade and the magnolids and reflect the basic features of the earliest angiosperms. Informally, however, it remains useful to compare the highly distinctive growth forms of the monocot clade with the more general growth forms of the remaining angiosperm clades, which we can refer to as the dicotyledonous grade. We will devote the final chapter of the book to the very distinctive monocots, which probably branched off somewhere between the magnolids and eudicots.

Dicots in this broad sense share some similarities with gymnosperms. They begin life as seedlings with two embryonic leaves, or cotyledons (Fig. 8.2), and to a large extent are woody, with permanent (axial) woody taproots or branching root systems. Their young stems have vascular tissues arranged in a distinct ring of bundles (Fig. 8.3), and their wood is formed as in gymnosperms, through a cylindrical vascular cambium that develops between the bundles (see Chapter 4, Fig. 4.16). Dicots are by no means all woody, however. As they diversified, a great variety of herbaceous forms also evolved, beginning with the archaic water lilies.

Monocots, on the other hand, begin as seedlings with a single elongate cotyledon, lack woody growth altogether, and have fibrous systems of replaceable (adventitious) roots that emerge from the stems. Though fundamentally herbs, there are some, such as palms and bamboos, with gigantic fibrous stems that are remarkably treelike.

Aside from possessing flowers, dicotyledons differ from most gymnosperms, as well as from monocots, in their distinctive and highly plastic leaves—broad

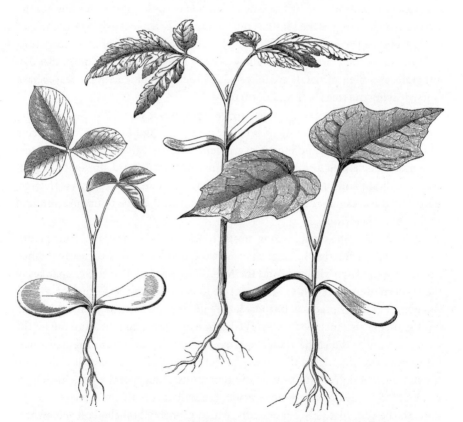

Figure 8.2 Dicotyledons begin life as symmetrical seedlings with two cotyledons, a small shoot apex between them, and a strong primary root. In most dicots, the primary root will continue to develop into a strong, branching, and often woody root system. Drawings from Kerner & Oliver 1895.

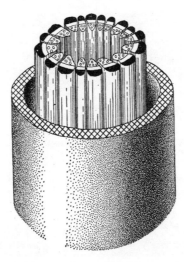

Figure 8.3 In the young stems of dicotyledonous plants, the vascular tissues are arranged in a ring of distinct bundles, with xylem tissues occupying the inner part of each bundle and phloem tissues and fibers in the outer part. The vascular cambium will develop as a thin cylinder running through each bundle between xylem and phloem. Drawings from Brown 1935.

leaves with a complex, netted pattern of veins that develop rapidly into quite varied shapes and are often deciduous after one season's use. Monocot leaves, on the other hand, elongate from the base and are typically sword- or paddle-shaped with parallel veins. Grasses, palms, lilies, and orchids are familiar monocots.

The contrasting sets of architectural features in dicots/gymnosperms and monocots signify rather different approaches to growth and survival. The adaptations of dicots favor above-ground shoot development, with branching into trees and shrubs, while those of monocots favor development of underground stem systems, especially during juvenile stages, and mostly herbaceous growth forms with simple above-ground shoots.

Vessels and the new wood of the dicots

The majority of dicotyledonous plants have a vascular cambium essentially like that of the gymnosperms, but one that produces a much greater variety of tissues. The wood of gymnosperms is relatively uniform, consisting of tracheids, thin plates of parenchyma (rays), and sometimes strands of resin-secreting cells. Wood in dicotyledonous plants has progressively become more complex, with mixes of tracheids, wider conducting tubes called vessels, fibers, and parenchyma (Fig. 8.4). So varied are the different mixtures and arrangements of these cells, that wood anatomists can identify samples of wood to the genus and often to a particular species. A medical examiner or forensic botanist might conclude for

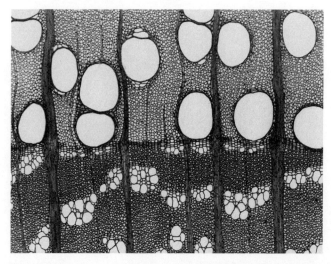

Figure 8.4 Dicotyledonous wood contains vessels of different sizes, fibers, tracheids, and parenchyma. Depicted here is *Celtis occidentalis* of the cannabis family, with large, round vessels, small fiber-tracheids, and parenchyma rays (dark vertical bands).

example that a murder victim was killed by a piece of wood from a maple species found only in a particular forest in northwestern Georgia.

Vessels are specialized, wide, water-conducting tubes in the xylem. They consist of a number of open-ended cells, called vessel elements, stacked together to form a sort of "super tracheid." Being longer and wider, vessels conduct water more rapidly through a tree than do a series of narrow tracheids. Eudicot trees of the temperate zone, such as oaks, maples, and others, take advantage of the rapid conduction of water through wide vessels produced in the early spring, when water is abundant and needed for rapid expansion of leaves and flowers.

There is a disadvantage to the wider vessels, however: they are more prone to disruption of their water column in times of dryness. When water is evaporating from the top of a tree more rapidly than can be supplied by the roots, tension increases in the columns of water running through the vessels. These continuous streams of water can then "snap," pulling away in both directions. This empties the vessels of water, and it is very difficult to reverse. This is the process of cavitation. Narrower vessels or tracheids typically surround the large vessels or are produced in the later wood growth of the summer, and these continue to conduct water more slowly if some of the wider vessels are cavitated.

In colder or drier climates, gymnosperms such as pines, spruces, and firs predominate. Their uniform wood of tracheids conducts water adequately for the slower growth of these plants and is less prone to cavitation. In the magnolid family Winteraceae, the wood lacks vessels, and the family was considered to have originated before vessels evolved. A new hypothesis has come out recently, however. The Winteraceae are related to other magnolids, such as the family

Canellaceae, which have ordinary vessels. It is now thought that the ancestors of the Winteraceae had vessels but lost them as an adaptation to colder climates, and they now have only tracheids (Feild et al. 2002). In this scenario, the ancient genus *Amborella* is the only living angiosperm with primitively vessel-less wood.

Other members of the ANITA grade and Magnolids have narrow vessels with slanted end walls with ladderlike plates of crossbars. This hints that vessel elements evolved from tracheids by gradual removal of the "membrane" (thin wall material) of pores at the ends of the cells. The more specialized vessels of eudicots are generally wider and have more horizontal end walls with few or no crossbars. Wide vessels are weaker than narrow tracheids, providing less physical support for the weight of a tree, so wood with vessels typically has masses of dense fibers to compensate (Fig. 8.3). This of course varies, and fast-growing, short-lived trees like balsa have much lighter wood.

Annual and biennial herbs—miniature trees?

It may seem strange to call a carrot a tree. The carrot family (Apiaceae) is a highly specialized eudicot family of herbs. Herbs are soft-bodied plants, with vegetation that typically lasts for only a single growing season. Carrots, radishes, and dandelions (Fig. 8.5) are herbs with the upright orientation and radial symmetry of a tree, but without the woody trunk. Some do have secondary growth in their

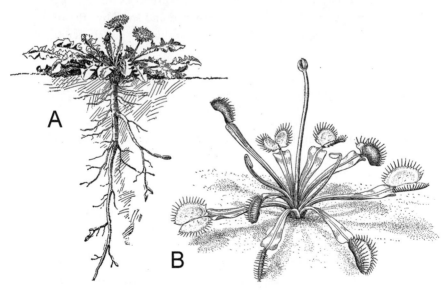

Figure 8.5 The dandelion (*Taraxacum officianale*; A) and the Venus flytrap (*Dionaea muscipula*; B) have very condensed stems, a long taproot, and leaves arranged in a rosette. Elongate stems are produced only to raise flowers above the ground. Drawings from Ganong 1916 (A) and Brown 1935 (B).

taproots, however. The roots increase in thickness, not by producing wood, but instead by adding layers of specialized food storage tissue. These herbs might therefore be imagined as a highly simplified trees.

These familiar tap-rooted herbs and many others, such as sundews and Venus flytraps, produce no stem above ground, at least during the first season. Their leaves form a compact, roundish mound at ground level, an arrangement known as a rosette (Fig. 8.5, also see below, Figs. 8.8, 8.19, and 8.20). Such plants are also referred to as acaulescent ("without a stem"), though this is a misleading term. All plants have a stem, even if it is just a stub that connects roots to leaves. During the appropriate season, most acaulescent plants form flower-bearing stems (inflorescences) and thus become "caulescent." In general, flowers must be raised above ground for effective pollination and seed dispersal.

Carrots, like many other root vegetables, are biennials, plants that live for two seasons. During the first season, nutrients are stored in the taproot, in preparation for flowering and seed production in the next season. We harvest the carrots, beets, radishes, and other roots at the end of the first season, appropriating the food resources for our own purposes and cheating the plants out the opportunity to reproduce.

Perhaps an 8-foot tall sunflower plant with its giant flower head is easier to imagine as a miniature, short-lived tree (Fig. 8.6). The sunflower is an annual herb, a plant that lives for just one growing season, but because of its upward growth and size, the sunflower stem requires strengthening tissue to hold itself up. This is supplied by a vascular cambium, which produces thin strips of wood. Wild sunflower species also tend to be branched and shrubby. The difference between a sunflower plant and a tree is that the sunflower's growth is limited to a single growing season.

Clonal trees

Dicotyledonous trees, shrubs, carrots, and sunflowers are solitary, upright plants with a strong central root-stem axis and a rough radial symmetry. Some dicots, however, have evolved means of spreading clonally like seedless vascular plants and bryophytes. Clonal plants form colonies of genetically identical, rooted units. The key to clonal growth is the ability to form adventitious roots. These are roots that arise from stems, making a new connection with the soil. We've seen this before in ferns and other seedless plants with horizontally spreading rhizomes. Clonal growth occurs sporadically among herbaceous dicots and is nearly universal among monocots, providing a means to occupy a large area of soil surface rather than form a single massive crown in the air.

Clonal growth is relatively rare in woody plants and virtually absent among living gymnosperms. Trees are usually anchored to a fixed spot of earth. The leafy

Figure 8.6 Sunflowers (*Helianthus* species) are annual or perennial herbs with elongate, leafy shoot systems that are strengthened by small amounts of secondary growth.

crowns and root systems of woody plants are balanced around a central trunk or around the point where the multiple trunks of a shrub merge into a common root system. There are a few woody plants, however, that have found ways expand into colonies of multiple trunks.

One way is the "top down" approach found in the banyan trees of the genus *Ficus* (Moraceae). From the broad, spreading branches of a banyan tree, adventitious roots emerge, extend to the ground, and take root in the soil. These dangling roots begin to thicken via secondary growth and become new trunks (Fig. 8.7). With this added support, the horizontal branches of the tree can extend further outward. More adventitious roots can then make more trunks, and the tree can expand into an ever-widening colony. The largest known banyan tree, a specimen of *Ficus benghalensis* in India, covers a little over 19,000 square meters. The red mangroves (genus *Rhizophora*) have a similar habit of producing stilt roots that support the leafy crowns above salty mud flats and allow them to spread laterally to some extent.

There are some 850 species of *Ficus* distributed throughout the tropics, plus the edible fig tree of the Mediterranean region. Some species of *Ficus* have remodeled adventitious roots adapted for even more bizarre applications. In the strangler figs, seeds germinate on the branch of another tree. The seedling sends down

Figure 8.7 The banyan tree (*Ficus benghalensis*) is unusual among woody plants for its habit of developing new trunks from dangling adventitious roots. Drawing from Gray 1879.

adventitious roots that wind their way down the trunk of the host tree, crisscrossing one another, and eventually fuse together. After a number of years of secondary growth, the roots have merged into a continuous hollow trunk that prevents further secondary growth of the host tree, eventually killing it. By the time the host trunk rots away, the hollow strangler fig trunk is strong enough to stand on its own.

A second way for woody plants to form a colony is from the "bottom up." The aspen trees of the Rocky Mountains, *Populus tremuloides*, form expanding groves by sprouting new stems from adventitious buds on shallow roots. The aspens on a Rocky Mountain slope often appear in the fall as patches of slightly different color, as each clone responds differently to the frosty cues of oncoming winter. The oldest colony, known as Pando, occurs in Fishlake National Forest in Utah. It is generally estimated to be 80,000, but possibly 1,000,000, years old, with 47,000 trunks occupying 43 hectares (Mitton and Grant 1996). Clonal woody plants can clearly live much longer than single-trunked trees.

Other trees that sprout new trunks from adventitious buds on their roots include the paper mulberry trees (*Broussonetia papyrifera*). Widely distributed in the islands of the South Pacific, the fibrous inner bark of this tree is beaten by local people into a paper or fabric (tapa cloth). Interestingly, *Brousonetia* is in the same family as *Ficus* (Moraceae). Other long-lived woody plants plant with the same habit include species of *Eucalyptus* in Australia, and even some conifers (*Lagarostrobos franklinii*, Podocarpaceae, in Tasmania; and *Picea abies*, Pinaceae, in Sweden). The eudicot creosote bush (*Larrea tridentata*, Zygophyllaceae) of the North American desert splits into separate woody clumps, each sprouting new stems and roots, as it ages. These clumps spread outward in a rough circle over time, and may live 11,000 years or more (Vasek 1980).

Many trees form adventitious buds at their base, not to create new separate units but to replace a main trunk that has been destroyed. I have seen this occur in kapok and silk floss trees in central Florida that were frozen to the ground in the late 1970s. After some 30 years, the trunk and crown of both trees now exceed their original dimensions.

Perennial herbs

Perennial herbs are plants that remain alive below ground year after year, though their foliage dies off at the end of each growing season. Most are so adapted as a means of surviving harsh winter or drought conditions, but there are also tropical perennial herbs that grow more-or-less continuously. The essential commonality is that perennial herbs never form permanent woody stems above ground. Though some may simply sit on a woody root crown, most have specialized root-bearing stems that can spread horizontally and are therefore clonal in nature.

A common adaptation among eudicots that superbly supports a clonal lifestyle is the stolon (Fig. 8.8). This is a slender horizontal stem with widely spaced nodes adapted for establishing new plants far enough away from the parent plant so as to not compete with it directly for root or leaf space. The strawberry (*Fragaria*) is the best known example of a plant that spreads by stolons. The individual strawberry plants are simple rosettes, like carrots, but without the swollen taproot. Stolons develop from specialized axillary buds. They extend horizontally, with greatly exaggerated internodes, touching the ground at intervals to form new rosette plants identical to the parent plant. The elongation of the stolons is due to a type of intercalary growth, similar to what occurs in the sporangial stalks of mosses and the vegetative shoots of horsetails.

The strawberry stolon forms above ground and is often referred to as a runner. Many stolons form below ground. The stolons of a potato plant plunge obliquely through the soil for a short distance, and then swell at the tip to form the food-filled potatoes, which are technically tubers. A tuber is a modified stem, which can be recognized by the many buds (eyes) on its surface.

The sweet potato is a member of the morning glory family (Convolvulaceae) and actually more of a ground-dwelling vine. It does not form rosettes like the potato or strawberry. Instead, the well-spaced nodes bear a single fully formed leaf, and if they touch ground, adventitious roots. Some of the adventitious roots swell into tuberous storage roots. The distinction between ground-covering vines and true stolons is thus slight and arbitrary.

Figure 8.8 Many rosette herbs form stolons that establish genetically identical daughter plants some distance from the parent plant. Pictured is *Saxifraga flagillaris*. Drawing from Kerner & Oliver 1895.

Highly successful colonizers of marshy ground include the pennyworts (*Hydrocotyle* spp., Araliaceae), which spread rapidly via slender stolons. Each node produces a single umbrellalike leaf, a cluster of adventitious roots, and an axillary bud that can branch off at an angle, expanding the colony in another direction. As lawn weeds, pennyworts are difficult to eradicate, because a single overlooked node can regenerate an entire new colony. As we'll see below, some pennyworts can even behave like water lilies when submerged.

The ability to develop a short rosette shoot and elongate stolons from the same plant illustrates the plasticity of the angiosperm body. Huge differences can result from whether or not internodes elongate or how much they elongate, and the flexible control of this process is one of the keys to angiosperm diversity. Long internodes result in a vine or stolon, while extremely short ones result in a rosette plant like a strawberry.

Short and long shoots result from regulation of localized cell division and growth by hormone systems, controlled primarily by gibberellins. This class of plant hormone was discovered in Japan as a result of fungal infections that caused "foolish seedling" disease in rice—pale, excessively elongate leaves and stunted root systems. The fungus produces a substance essentially like the gibberellins later found to be natural hormones in plants. Artificially applied gibberellin can cause dwarf varieties of corn to grow to the height of ordinary corn, and it can cause cabbages to string out like bean stalks (Fig. 8.9). A number of hormones control the development of the plant body, but none is more dramatic in its effects than gibberellin.

By definition, a rhizome is a horizontal stem bearing adventitious roots on its lower surface, with leaves and shoots on the upper surface. Rhizomes are typically shorter and thicker than stolons and often function for food storage. They consist of a continuous, occasionally branching plant axis, producing a progression of leaves or upright shoots, but not separate individual plants, per se, except through fragmentation. A stolon can, however, be considered a specialized form of rhizome, and there is no hard distinction between the two forms.

Most familiar examples of thick rhizomes, such as those of ginger or iris, are monocots, but there are a few among the dicotyledonous plants. The term paleoherb has been used for perennial herbs in the Nymphaeles (water lilies) and in the magnolid Piperales and Aristolochiales, but we now know that these are not closely related to one another. Paleoherbs typically have simple, thick rhizomes, as do eudicots like *Gunnera*, certain species of *Begonia*, and clumping members of the Ranunculaceae and Araliaceae (Fig. 8.10). Monocots most likely evolved from a paleoherb ancestor related to the Chloranthaceae, which sometimes is included in the Magnoliales.

Perennial herbs also include plants with tuberous roots, such as *Dahlias* and sweet potatoes. In *Dahlias*, a cluster of thick food-filled roots develops beneath a short stem crown from which buds develop into herbaceous leafy shoots and flowers each season (Fig. 8.11). The sweet potato, as mentioned above, forms tuberous roots at intervals along its sprawling stems. Some species of peony are shrubby, but others form fresh leafy shoots each year from a woody root crown just below the soil surface.

Many perennial herbs have swollen stems just below the surface that are referred to as tubers or sometimes as corms. In the tuberous begonias, and in the unrelated gloxinias, the roundish tuber can produce a cluster of fresh shoots from buds on its upper surface after a period of dormancy.

Vines

We think of vines as climbers, decorating fences, trellises, and arbors in our gardens. In nature, they clamber up through trees to capture a share of the brightly lit canopy (Fig. 8.12). Many vines, however, are equally at home sprawling across the ground, as is the sweet potato vine mentioned above.

Figure 8.9 Professor Sylvan Wittwer examines cabbage plants treated with gibberellin. Photograph courtesy Mrs. Wittwer.

Figure 8.10 Anemone nemorosum, a member of the Ranunculaceae, creeps below ground by means of a horizontal rhizome, before turning upward to form an upright shoot with leaves and flowers. Drawing from Kerner & Oliver 1895.

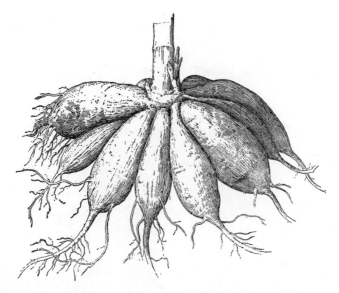

Figure 8.11 The common ornamental plant, *Dahlia*, has a cluster of tuberous roots attached to a short stem section that contains buds for the next season's growth. Drawing from Brown 1935.

Vines typically have long internodes adapted for rapid growth, either horizontally or vertically. If a vine sprouts roots at its nodes, it, too, is clonal. Some can form dense ground covers over abandoned fields and roads, and when chancing upon a suitable tree or shrubby border, climb up the side to the canopy. Once they begin upward growth, vines may become woody and the equivalent of trees.

In their natural habitat, such vines play the important role of covering over bare ground, retarding erosion, and paving the way to reforestation. Such aggressive vines are mostly denizens of the tropics and subtropics, though blackberry brambles and wild grapevine play similar roles in cooler climates. Some become horrific weeds. The infamous kudzu vine (*Pueraria lobata*, Fabaceae) forms vast, suffocating colonies that kill trees and bury cars and even whole houses in the southeastern United States. Air potato vine (*Dioscorea bulbifera*, Dioscoreaceae), skunk vine (*Paederia foetida*, Rubiaceae), and flame vine (*Pyrostegia ignea*, Bignoniaceae) have similar habits, if not quite so aggressive, and are weeds in Florida.

The leaves of dicotyledonous plants

The leaves of dicotyledonous plants are extraordinarily variable in shape, though based on a fundamental plan that includes a narrow petiole and a broad, net-veined blade.

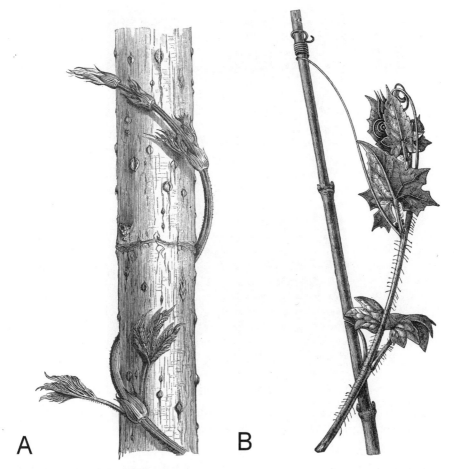

Figure 8.12 Vines may twine by their touch-sensitive stems, as in *Humulus* (A), or by tendrils, as in *Bryonia* (B). Drawings from Kerner & Oliver 1895.

The main vein, or midrib, emerges from the stalk of the leaf, the petiole, and typically runs through the flat blade to its tip. It may also split at the base into several equal ribs, especially in very broad or divided leaves. Smaller veins can be seen branching off of the main veins and may end as teeth at the leaf margin or loop back to reconnect with one another. Still smaller veins form bridges between those veins, resulting in a hierarchy of subdivided patches, each bordered by a ring of vascular tissue. Within each of these smallest patches, one or more tiny veinlets enter at right angles and end blindly amongst the photosynthetic cells (Fig. 8.13). This pattern of branching and reconnecting veins is referred to as reticulate or net venation.

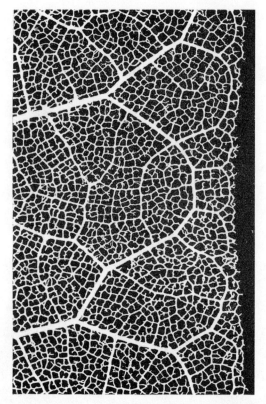

Figure 8.13 The intricate hierarchy of veins in a dicotyledonous leaf is due to uniform expansion of the young leaf. Drawing from Coulter et al. 1910.

How does this complex pattern arise? Dicotyledonous leaves typically form their shape in miniature, then expand in all directions to reach full size (Fig. 8.14). It is like drawing a picture on an empty balloon and then watching it expand as the balloon is inflated. The generalized expansion of the leaf, with dividing, expanding cells scattered throughout, is referred to as a plate meristem. The first veins to appear in tiny new leaves are the main veins entering from the petiole. As the leaf expands, new branch veins appear spontaneously between the earlier veins. This process repeats continuously until the leaf reaches full size, resulting in the characteristic reticulate network.

This sort of development and vein network is found in most dicotyledonous plants, and therefore evolved among the earliest angiosperms. It is rare in living gymnosperms, but a similar pattern is found in the leaves of the gymnosperm genus *Gnetum*, in the enigmatic Gnetales. Similar patterns can also be seen in some ferns, although most fern fronds develop fundamentally from branching apical meristems, more like a shoot system. Monocots, the subject of the next chapter, developed a completely different leaf architecture, one based on growth

Figure 8.14 In flowering plants other than monocots, leaves form in miniature within buds then expand uniformly to their full size. Pictured is a species of *Liquidambar*.

from the base, which maintains separate, parallel veins running the length of the leaf. Some specialized monocot leaves, however, have come to resemble the dicotyledonous pattern.

Leaves come in different sizes. Small, simple leaves may be produced in profusion on finely branched twig systems, or fewer large leaves may be produced on stout, sparsely branched twigs. These larger leaves are usually subdivided into lobes or separate leaflets (compound leaf) (Fig. 8.15). Divided leaves are less prone to damage by the wind, allow for greater movement of diffused gases into, out of, and around the leaf for enhancing gas exchange, and are less prone to overheating in the midday sun.

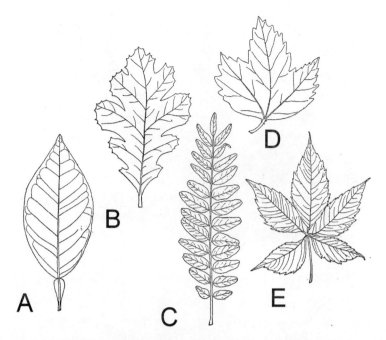

Figure 8.15 Dicotyledonous leaves may be simple, as in the orange (*Citrus*; A); pinnately lobed, as in the oak (*Quercus*; B); pinnately compound, as in the locust (*Robinia*; C); palmately lobed, as in the highbush "cranberry" (*Viburnum*; D); or palmately compound, as in the Virginia creeper (*Parthenocissus*; E). Drawings from Ganong 1916.

Taken to an extreme, this trend results in rosette trees, such as papayas, that seldom branch and have a massive cluster of large, divided leaves emanating from the terminal bud (Fig. 8.16). A rosette tree is a more massive version of rosette herbs like carrots and strawberries. Cycads and tree ferns exploited this growth form hundreds of millions of years before any angiosperms, as have more recent monocots such as palms. A technical name for such growth forms is pachycaul (thick stem), for the trunk that supports the massive terminal crown of compound leaves seems disproportionately thick.

Bracts, spines, and tendrils

Leaves can be modified in extraordinary ways for specialized functions. A bract is any kind of leaf modified in size, shape, or color, but still basically flat. Colored bracts may help animal pollinators locate flowers. What appear to be giant red petals in poinsettia blossoms are actually colored bracts. Stiff, fibrous bracts may surround dormant buds during winter or protect soft developing shoots. Of course, in preangiosperms, even the parts of the flower evolved from specialized leaves.

Leaves, or parts of leaves, that are reduced to hard, sharp points are called spines, not to be confused with thorns, which are short modified stems, or prickles, which are outgrowths of the surface tissues of a stem (Fig. 8.17). The "thorns" of a rose bush are actually prickles. Finally, either leaves, parts of leaves, or special side stems of vines can be modified into touch-sensitive tendrils, which wrap around slender twigs or wire fencing in support of upward growth (see Fig. 8.13).

Deciduous leaves

Leaves themselves are expendable and may be shed at the onset of winter or seasonal drought. Whether or not a tree is evergreen or deciduous, however, depends on a number of "economic" factors. There is a certain expense in creating new leaves: including

Figure 8.16 A papaya tree, like a tree fern, cycad, or palm, has a single thick trunk and large compound leaves arising from a massive terminal bud.

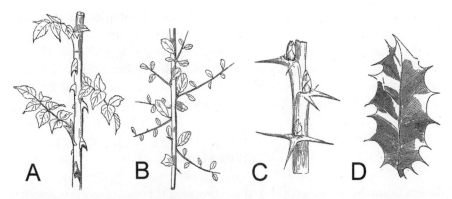

Figure 8.17 Sharp, defensive structures have evolved many times and in many ways. A. prickles, as in the rose, are outgrowths of the surface of a stem; B. true thorns are modified stems; C. spines are modified leaves or D. portions of leaves. Drawings from LeMaout & Decaisne 1876.

water, mineral nutrients, and organic energy. If the growing season is relatively long, the expenses can be recovered, and deciduous trees are common. Deciduous leaves are typically broad, thin, and relatively soft, in contrast to those adapted for winter survival, and so are optimally shaped for photosynthesis. Evergreen trees must invest relatively heavily in fortifying their leaves against damage from freezing, desiccation, and herbivores, and such leaves tend to be rather small and tough.

Deciduous trees dominate in the temperate forests of eastern North America and Eurasia, as well in the tropics where there is a long rainy season alternating with a prolonged dry season. When rivers run through drier climates, trees along them are able to function throughout the summer even in the absence of rain, and so can afford to replace their leaves each year.

In the humid tropics, trees can be active year-round, so they are generally evergreen. Individual trees, however, may drop all their leaves before developing a new flush. This may help reduce the number of insect pests, by periodically forcing the insects to relocate. Insects are typically highly specific about which plants they feed upon, and in tropical rain forests individuals of a particular species are usually far apart, and mixed with many other species. This greatly reduces the spread of insects from tree to tree. Even the devastating rampage of millions of army tent worms in a temperate forest can be tolerated, as the trees can remain leafless long enough for the insect swarms to move on.

In relatively hostile environments with short growing seasons, trees and shrubs are also most likely to be evergreen, because the cost of producing new leaves cannot be recovered during the limited time available for photosynthesis. Also, with leaves always on the trees, photosynthesis can start up quickly during an unexpected warm or rainy spell. The vast boreal forests of Siberia and North America are therefore dominated by evergreen pines, spruces, and other conifers, as are high mountain forests. For similar reasons, areas of hot, dry summers and wet winters, like the chaparral of California and the Mediterranean region, are also likely to host primarily evergreen trees and shrubs.

Xerophytes

After sand dunes, the most recognizable icon of the desert is the cactus (Fig. 8.18A). These plump, succulent (water-filled) stems have no leaves, or to be more accurate, their leaves have been converted into spines. Photosynthesis occurs in tissues just below the epidermis of the stems. Leaves, by nature, lose a great deal of water as they exchange gasses, so getting rid of them and turning the photosynthetic function over to the more water-conserving stems, is a prime option for plants living in extremely dry environments. Swollen stem succulents (Fig. 8.18), such as cacti are one form of xerophyte ("dry plant").

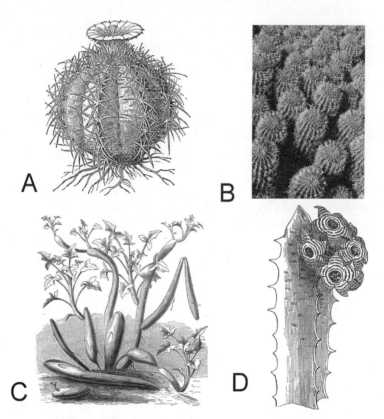

Figure 8.18 Succulent stems have evolved multiple times, most notably in the cactus family (A; *Echinocereus*), euphorbia family (B; *Euphorbia*), sunflower family (C; *Kleinia*), and milkweed family (D; *Stapelia*). Drawings from Ganong 1916 (A), Kerner & Oliver 1895 (C), and LeMaout & Decaisne 1876 (D).

The thick stems of cacti have a lower surface-to-volume ratio, and therefore lose less water than thin leaves. They are also covered with a thick waxy cuticle, and their stomata are typically closed during the day. In order to supply sufficient carbon dioxide for photosynthesis, the stomata open during the night, and as the CO_2 diffuses in, it is incorporated into 4-carbon molecules of malate. During the day, the CO_2 is released into the photosynthetic cells and used to make sugar with energy supplied by the light reactions.

The metabolic pathway for capturing, storing, and releasing CO_2 in this fashion is called crassulacean acid metabolism (CAM). During the long dry spells characteristic of the desert, stomata may be closed altogether, but the cacti can remain minimally active by recirculating carbon dioxide and oxygen internally. Obviously they do not grow very much at such times, but they stay alive.

True cacti (Cactaceae) are native only to the Americas, but similar-looking stem succulents (Fig. 8.18B, C, D) have evolved independently on other continents from other plant families, providing striking examples of convergent evolution. In Africa and Madagascar, succulent species of *Euphorbia* and stapeliads (*Stapelia* and related genera in the Apocynaceae) abound. Some *Senecios* (Asteraceae) have also become cactus-like. Some *Euphorbias*, *Pelargoniums* (Geraniaceae), the grape relative *Cissus quadrangularis* (Vitaceae), *Pachypodiums* (Apocynaceae), and *Adenia* (Passifloraceae) have a two-tiered strategy. During the brief rainy season, they employ ordinary-looking leaves for photosynthesis, but drop them at the beginning of the dry season. Then the greenish, water-filled stems function like cacti and continue to photosynthesize.

Other plants have turned their leaves, instead of their stems, into thick water-storage organs. The eudicot family Crassulaceae consists almost entirely of leaf succulents (and incidentally, lent their name to crassulacean acid metabolism). Most members of the Aizoaceae and Portulacaceae also have succulent leaves, as do some members of the sunflower family (Asteraceae). Still others survive the dry climates as ordinary shrubs by just toughing it out during dry weather. The dominant shrub of North American deserts, the creosote bush (*Larrea*), does this with thin, highly fibrous leaves (sclerophylls) that are resistant to or tolerant of desiccation. Most evergreen trees and shrubs in the summer-dry woodlands of the Mediterranean chaparral have similar fibrous leaves.

Carnivorous plants

Nothing illustrates the plasticity of the dicot leaf better than carnivorous plants. In the many kinds of plants that have taken up the consumption of small animals, it is the leaves that have been turned into traps and stomachs. Plants do not become carnivorous in the same sense as animals. They still depend on photosynthesis for their energy needs. Carnivory is primarily for obtaining mineral nutrients, and appears to have evolved in habitats where soils are poor in minerals.

Carnivory has probably evolved independently at least six times among plants, resulting in a dozen or more genera and 630 species. Fully carnivorous plants attract and trap prey, produce digestive enzymes, and absorb the nutrients that are released. How do plants embark down the seemingly unlikely road to carnivory? Some plants have only some of these characteristics, and they give us some clues as to how this habit might evolve.

There seem to be at least two starting points: water tanks and sticky hairs. Though most true carnivorous plants are eudicots, the monocot family of bromeliads shows us how water reservoirs can be turned into digestive vats. Epiphytic bromeliads routinely collect water in tanks formed by their tightly overlapping leaves. Debris falls into these tanks, and aquatic insects, frogs, and even fish have been known to live and breed in them. No doubt some nutrients from the decayed waste products of these animals can be obtained by the plants as they absorb water from their tanks. In *Borrichia reducta*, however, there have been some specific adaptations to increase their carnivorous nature: a slippery coating on the leaves causes insects to fall in, and they are then digested by bacteria. The plants themselves produce no digestive enzymes.

In eudicots, slippery, water-filled traps form from individual leaves that have been highly modified and hollowed out (Fig. 8.19). Pitcher traps typically have attractive color patterns and sometimes nectar-like secretions to lure the animals in. Downward-pointing hairs hasten the slide and prevent the insects from crawling back out. Most also have digestive enzymes to hasten the decay of the prey, and a few are believed to secrete drugs that impede the insects' ability to crawl out.

One can only imagine that slight deformations of their ancestral leaves first served to collect water and that these were progressively transformed into elaborate insect traps. Some pitcher plants rely on bacteria for digestion of their prey, while others secrete digestive enzymes. The latter step qualifies them as full-fledged carnivorous plants. Pitcher plants have evolved independently in three different families of eudicots. *Sarracenia*, *Darlingtonia*, and *Heliamphora* are in the Sarraceniaceae, while *Nepenthes* and *Cephalotus* are each in their own small family.

The other major path to carnivory was through the use of sticky hairs as a type of fly paper. In the South African shrub *Roridula*, stiff, glandular hairs cover the plant. The sticky secretions trap the many insects that wander onto the plant. The insects are not digested, however, but left to decay. This is believed to be a defensive adaptation to keep the insects from feeding on the plants. It is possible that the plants benefit indirectly if nutrients from the decaying insects wash into the soil. The next step would be to evolve digestive enzymes and to absorb dissolved minerals through the surface of the leaves. *Roridula* is not directly related to any

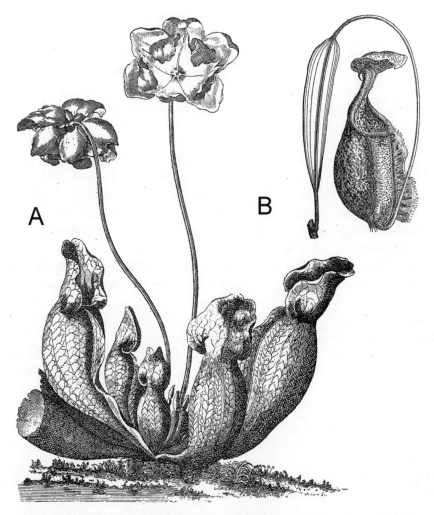

Figure 8.19 Pitcher trap leaves have evolved several times. In *Sarracenia* (A), pitchers arise from a compressed stem at ground level, while in *Nepenthes* (B), they form at the tips of leaves borne on vines. Drawings from Ganong 1916, attributed to Barton (A), and LeMaout & Decaisne 1876 (B).

other carnivorous plants, but it provides a model of adaptations that could set the stage for true carnivory.

A family that did exploit this potential to its fullest is the Droseraceae, which includes sundews, the Venus flytrap, and the aquatic *Aldrovanda*. Here, leaves with sticky hairs secrete digestive enzymes and absorb nutrients through their flat surfaces. Hundreds of species of sundews (*Drosera*) attest to the success of this strategy (Fig. 8.20A). Some of them fold around their prey after a capture, creating a loose digestive cavity, which perhaps enhances the efficiency of the digestive process. The seedlings of *Drosera* are at first extremely small, but their tiny leaves, too small to trap insects, are already secreting fluids and are pressed to the ground. There they are confluent with the watery channels of the wet soil. We can only speculate at this point that they may be catching protists.

In the Venus flytrap (*Dionaea*; see Fig. 8.5), the ability of the leaf to close is taken to a highly sophisticated level. The two halves of the leaf snap together through a rapid change in water pressure. The sticky hairs of their ancestors have disappeared, except for three that have been modified into triggers. When bumped twice, the trigger hairs release a nerve-like impulse that travels throughout the leaf, initiating the closing mechanism. *Aldrovanda* does something similar, but under water.

It is less clear how the bladderworts of the genus *Utricularia* (Lentibulariaceae) got their start. They function under water or in wet soil. The feeding leaves take on the form of tiny bladders (Fig. 8.21), which create a partial vacuum by pumping water out osmotically, pulling the elastic walls of the chamber inward. The chamber is sealed by a small door, which opens inward when tiny invertebrates bump against the surrounding hairs. The vacuum sucks the prey inside where they are digested. Whether the trigger hairs evolved from sticky hairs, as they did in the Venus flytrap, is not known, but the related butterworts (*Lentibularia*) have much simpler traps consisting of leaves covered with very short glands that make the entire upper surface sticky.

The most recently verified carnivorous plants are in the genus *Genlisea*, also in the Lentibulariaceae. Suspected of carnivory by Charles Darwin, scientific studies reported in 1998 finally demonstrated the capture of protists by the plant's

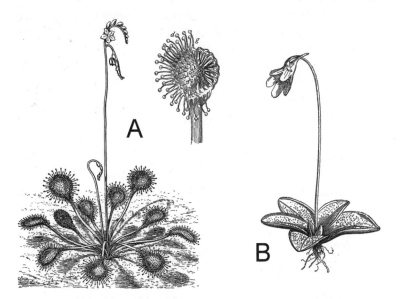

Figure 8.20 Sundews (*Drosera*; A) and butterworts (*Pinguicula*; B) have sticky hairs for trapping prey, though those of the butterworts are tiny. Drawings from Brown 1935 and Coulter 1910 (A), and LeMaout & Decaisne 1876 (B).

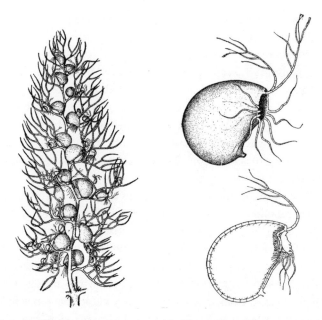

Figure 8.21 In the bladderwort, *Utricularia*, bladder traps are borne on highly divided underwater leaves. Trigger hairs at the opening of the trap cause the sudden influx of water when touched, pulling the prey item inside. Drawings from Brown 1935.

highly specialized underground leaves (Barthlott et al. 1998). These leaves are colorless, Y-shaped, and tubular, superficially resembling roots. Protists enter through slits along the sides of the tubes and are prevented from exiting by stiff hairs. The protists migrate toward the junction of the two arms, lured by a chemical attractant. There they are digested in a small chamber. These are generally referred to as lobster-pot traps.

Aquatic plants

The ancient water lilies (Nymphaeales) constitute a clade almost as old as the *Amborella* clade, and they are unusual among the ANITA grade for being entirely aquatic and nonwoody. All 74 species, in three families and six genera, are aquatic. The broad net-veined leaf blade of a water lily arises from a rhizome anchored in the mud. The petiole elongates through intercalary growth to bring the blade to the surface. Nothing else like them is found among the other ANITA clades, and among paleoherbs in the Magnolids one finds only marginally aquatic species, such as *Saururus* (Saururaceae). Similar aquatic growth forms have evolved independently in many unrelated families of Eudicots and monocots.

Though there are many woody plants tolerant of standing in water (red maples, pop ashes, and mangroves, to name only a few), the most varied and interesting aquatics are herbaceous. Many eudicot herbs, such as *Bidens* spp. in the sunflower family (Asteraceae), *Polygonum amphibium* (Polygonaceae), *Decodon* spp. (Lythraceae), and *Ludwigia* spp. (Onagraceae), enter the water stiffly like trees, with conventional leafy shoots above the water. These are apparently more recent aquatics, tolerant of standing in water but without any of the more iconic adaptations of aquatic plants.

More specialize aquatics typically take the form of rhizomes rooted in the mud, with often roundish leaves and flowers rising individually to the surface of the water, as in the fashion of water lilies (Fig. 8.22). A eudicot genus remarkably similar to a water lily is *Nymphoides* in the family Menyanthaceae. Though not quite so grand, these plants produce similar roundish lily pads (Fig. 8.22B). The lotuses (genus *Nelumbo*) (Fig. 8.22C) were long considered close relatives of the water lilies but are now known to be more closely related to sycamores (Platanaceae) in the eudicots. They, too, creep in mud under the water, periodically sending up rounded leaf blades on long petioles. Occasional large flowers, remarkably similar to those of the water lilies, rise above the water as well. *Trapa*, the water caltrop (Lythraceae), has been grown in China for hundreds of years for its edible seeds. It has more diamond-shaped floating leaves.

The genus *Hydrocotyle* in the Araliaceae is a remarkably adaptable colonizer, appearing as a weed in moist lawns but also stretching runners into fresh water, and if anchored underwater, sending rounded leaves upward 30 cm or more on elongating petioles. Roundish leaves appear to be the optimal shape for floating. Leaves that have to hold themselves up against the force of gravity, on the other hand, are either dissected into segments or have strong midribs running the length of elongate blades. The round shape of floating leaves distributes the photosynthetic area evenly around the central supply of the petiole.

Alternatively, aquatic plants may become totally submerged and take on a form resembling charophyte algae, with leaves clustered in whorls (circles) along slender stems and divided into narrow segments (e.g., *Cabomba*) (Fig. 8.23). Examples from the eudicots include *Ceratophyllum, Myriophyllum, Utricularia,* and *Hippuris*. Some *Cabombas* as well as aquatic species of *Ranunculus* produce two kinds of leaves: rounded, floating leaves similar to those of water lilies above water, and dissected leaves below water (Fig. 8.23A, B). Highly dissected leaves with very narrow segments are apparently an adaptation for allowing maximum water flow around and through the leaves, reducing resistance to currents and increasing the availability of carbon dioxide.

Flowers of all water lily and eudicot aquatics are lifted above water for ordinary aerial pollination. Special air-conducting tubes in the roots, petioles, and rhizomes provide a passage for oxygen-rich air needed by the roots anchored in oxygen-poor mud. Similar air-filled cells in the leaves provide flotation. Both kinds of tissues are referred to as "aerenchyma."

Figure 8.22 The leaves of aquatic plants that emerge to the surface tend to be large and roundish. Such growth forms have evolved a number of times, including *Nymphaea* (Nymphaeaceae; A), *Nymphoides* (Menyanthaceceae; B), *Nelumbo* (Nelumbonaceae; C), and in several other dicot and monocot families. Drawings from Masclef 1891 (A), Sturm 1796 (B), and Kerner & Oliver 1895 (C).

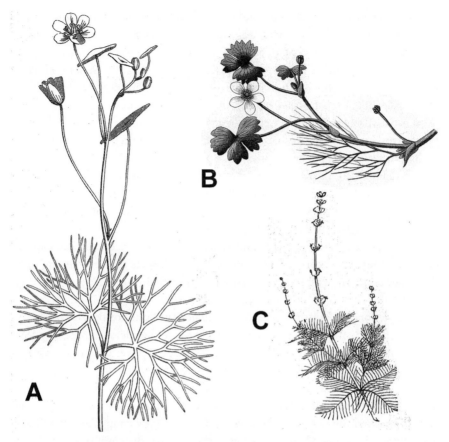

Figure 8.23 A. *Cabomba* has broad, roundish leaves, like other members of the Nymphaeales, but those that develop underwater are divided into narrow segments. B. *Ranunculus aquatilis* also produces normal leaves above water and dissected leaves under the water. C. The leaves of *Myriophyllum*, in the Haloragaceae, are all underwater and feather-like. Drawings from LeMaout & Decaisnes 1876 (A), Thomé 1885 (B), and Die Gartenlaube 1887 (C).

Secondary plant compounds—how plants defend themselves

Overlain on the diverse forms of flowerings plants is the invisible world of plant chemistry. Cyanide, anise, and caffeine to us are a poison, a flavoring, and a stimulant drug. To the plant world, however, they are the first line of defense against herbivores. They are just three out of thousands of secondary plant compounds produced by plants—chemicals that are often unique to particular species. There are so many because the organisms attacking them, primarily insects, are varied and capable of rapid evolution. As plants evolve new defensive compounds, insects almost as rapidly develop ways around them. Insects tend to specialize in particular plant foods, and so natural selection intensely favors toleration of the compounds produced by those plants. The result is each plant successfully deters all potential predators except for a few specialists.

A good example is the highly toxic oleander plant (*Nerium oleander*, Apocynaceae), which is fed upon in the southeastern United States by the "oleander caterpillar" (*Syntomeida epilais*) and little else. Interestingly, the oleander plant is native to the Middle East, while the oleander worm is native to the New World tropics. In its native habitat, this insect possibly fed on related plants, such as *Echites umbellata*, before becoming a pest on oleanders planted in Florida (McAuslane 2012). In southern Asia and Australia, oleander is fed upon by the common crow butterfly (*Euploea core*). Both kinds of insects are able to safely store the oleander toxins in special tissues. This makes them poisonous to most animals that might otherwise feed upon them.

Some trees, like the apple (*Malus*, Rosaceae), produce secondary compounds only after they have been attacked, not to kill the insects that are feeding upon them, but to attract wasps that will parasitize the plant-eating insects (Turlings et al. 1990).

Dicotyledonous plants provide the vast majority of our poisons, herbs, spices, drugs, and medicinal compounds, perhaps because their leaves are so exposed. Gymnosperms are not without their toxic arsenal, but fewer in number. Cycads are quite toxic, and important herbal drugs come from *Ginkgo biloba* and *Ephedra*. Ferns can be quite toxic as well. Monocots, as we'll see in the next chapter, rely more on buds buried underground or within overlapping leaf bases, and on rapid growth to avoid the ravages of herbivores. Grasses, for example, are mercilessly grazed upon by herds of mammals but survive by rapidly sprouting new leaves from underground rhizomes. Nevertheless, the monocots give us the varied medicinal and culinary compounds of onion, garlic, and their relatives (family Alliaceae), the steroids that are derived from *Dioscorea*, the vanilla extract that is derived from a species of orchid, and ginger, turmeric, and related herbs in the Zingiberaceae.

The dicotyledonous grade, though not a single taxonomic entity, is united by a fundamental but flexible form of leaf development and displays a great array of growth forms, including everything from giant woody trees to tiny aquatic herbs. In the final chapter, we'll explore the monocots, which are characterized by a much more restricted developmental plan, one that prohibits wood but excels at surviving underground. That restricted plan, however, has led to vast new possibilities and a group nearly as varied as the eudicots.

Figure 9.1 The rigors of the African savanna force a stark dichotomy between woody plants that rise above fire and herbivores, and perennial herbs that tunnel below. The grasses that dominate this habitat are monocots specialized for survival underground.
Photograph by Sandra Fenley, posted on Wikimedia Commons, licensed by Creative Commons.

9

The Monocots

The savannas of Africa are home to the most spectacular display of animal life to be seen anywhere on the planet. Vast herds of grazing animals traverse the continent following seasonal plant growth, and they in turn are pursued relentlessly by a variety of carnivores. The vegetation that supports this food chain is, however, rather simple and monotonous. Here and there we see a lone *Acacia* tree or small groves of them, sometimes another shrub or wildflower, but what dominates the scene is grass—mile after mile of grass (Fig. 9.1).

The savanna is a harsh environment for plant life. Tree and grass alike must endure intense grazing and sometimes flooding during the wet season. During the equally long dry season, bone-dry soil and frequent wildfires provide even greater challenges. *Acacia* and grass alike flourish among these rigors, but in radically different ways.

Acacias are woody plants that survive by stretching vertically in both directions, lifting foliage above grazers and fire, and sending long taproots toward deep water reserves. Their bark provides insulation against the relatively low heat of grass-fed ground fires. Like other trees, *Acacias* are fixed to a single spot of ground, with both crown and root systems balanced around their central axis in a rough radial symmetry.

Grasses survive the same hazards, not by lifting themselves above the many threats, but by burrowing beneath them. Leaves and flowering shoots are expendable and sacrificed in times of stress or attack, but the underground stems (rhizomes) persist indefinitely (Fig. 9.2). Provisioned with food reserves and insulated by a blanket of soil, grass rhizomes lie dormant through adversity, quickly sprouting new foliage when conditions are right.

The leaves of grass are sword shaped and sharp pointed, with parallel veins running through them. They lengthen exclusively through growth at the base, and this pushes the hardened, mature tips upward through the soil. If the blades of young grass leaves are burned or eaten (or mowed), basal growth may resume, pushing the stubs upward to replace the lost tissues.

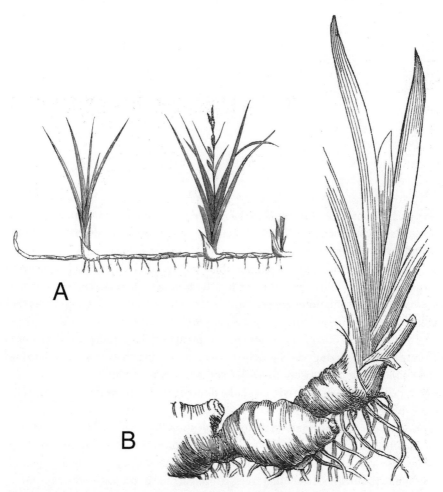

Figure 9.2 The basic stem in monocots is a rhizome similar to those found in ferns and other lower vascular plants. Each new segment begins horizontally, bearing adventitious roots and bracts, then turns upward into a leafy and/or reproductive shoot. New horizontal segments form as buds at the base of the upright shoots. In this sedge (A), the rhizome is elongate and adapted for maximum spread of the colony. In the bearded iris (B), the rhizomes are short and thick, adapted more for food storage than territorial expansion. Drawings from Thomé 1877 (A) and Gray 1879 (B).

Roots are also expendable, disintegrating along with older sections of the rhizome, and replaced at the growing tips. As the rhizomes extend horizontally, they branch frequently, forming extensive colonies of genetically identical units. Grasses are therefore also clonal, like many ferns and other seedless plants. Rhizomes and life as perennial herbs are key feature of the monocots.

The dichotomy between *Acacia* and grass is a theme nearly as old as plant life itself. The first trees rose above clonal bryophytes and seedless vascular plants some 360 million years ago, though probably in swampy conditions. Savannas similar to those of today existed 150 million years ago, where one would have seen dinosaurs grazing on fast-growing ferns beneath ancient pines and ginkgos. Grasses now dominate not only savannas but virtually all other wide-open, seasonally wet ecosystems around the world, resulting in a wide variety of "grasslands."

Grasses constitute a single family of plants, the Poaceae, but this is just one of about 82 families constituting the large and diverse clade of angiosperms known as monocots. At least some members of all these families exhibit the same horizontal, underground growth form characteristic of grasses, and most have sword-shaped leaves in at least their juvenile foliage. Though some, like palms, become trees, none have woody stems, and roots are always adventitious. The name given to this group is based on the fact that the embryo has but a single seedling leaf, or cotyledon. That cotyledon has the same general structure as the typical sword-shaped leaf.

In all, there are 2755 genera and over 60,000 species of monocots, adapted to a great variety of habitats, from arctic tundra to desert, tropical rain forest, and even shallow seabed. The grasses are one of the most advanced and certainly the most abundant of monocot families, but at the same time they exhibit most clearly the theme of clonal, underground stem systems and upward-penetrating leaves upon which all monocots are derived. The great diversity of monocots also include spring-flowering bulb plants, like tulips and daffodils, as well giant tropical herbs like bananas and gingers. They include epiphytic orchids and bromeliads, desert succulents like *Agave* and *Yucca*, as well as all manner of rooted, floating, and submerged aquatics. There are even vines and giant treelike monocots in the form of palms and screwpines (*Pandanus*).

The monocots are not quite as numerous as the eudicots, but have evolved a great diversity of distinctive growth forms. As clonal plants, they are a new cutting edge of plant evolution, following in the tradition of ferns and other early plants, but with new versatility and modes of reproduction. They dominate certain habitats: clearly the grasslands, but also marshy and aquatic habitats with vast swathes of cattails, sedges, and rushes. As bamboos, they have taken over certain forests in Asia, and as palms, they dominate tropical beaches, desert canyons, and swamp forests in New Guinea and the Amazon Basin. Finally, monocots in the form of sea grasses are the only vascular plants to have fully adapted to the marine environment, where they form extensive meadows in shallow coastal waters.

All of the unique growth forms of monocots center around their unusual leaves, which began with adaptations for rising from underground stems, but then took on many new shapes as they adapted to different habitats and above-ground growth forms.

The cotyledon

The single cotyledon of a monocot embryo has the same general form as the foliage leaves, with a long, narrow blade and a basal sheath surrounding the embryonic shoot. There is evidence that the single cotyledon came about through the fusion of the bases of two ancestral cotyledons, as it has two main veins that originate on opposite sides of the bud (Stebbins 1974).

As the embryo develops, growth of the cotyledon dominates, shifting the relatively tiny terminal bud to the side. Monocot cotyledons rarely see the light of day. During germination, the tip typically remains inside the seed to absorb nutrients from the endosperm, and basal growth of the cotyledon pushes the embryo out of the seed and generally deeper into the soil where there is more moisture (Fig. 9.3). This is an important adaptation in areas of uncertain or seasonal rainfall, such as a savanna. Leaf blades then arise one by one from this underground refuge, their basal sheaths adding layers of protection around the bud.

The monocot leaf

The basic, sword-shaped monocot leaf is roughly the same width throughout its length until it tapers to a point at the tip (Fig. 9.4). The principle veins of the leaf emerge individually from the stem and run parallel to one another through the length of the leaf, converging at the tip. Thin crossveins form between the main veins, forming a two-dimensional network. Cell division and expansion are confined to a narrow band at the base of the blade called a basal intercalary meristem. So as the sword-shaped leaf grows, new tissues at the base push older tissues upward (Fig. 9.5). New conducting cells are likewise added to each vascular bundle at the base of the blade.

The base of each leaf encircles the entire stem, forming a leaf sheath. The sheath forms a protective pocket around the apical meristem and younger developing leaves. A very young monocot leaf resembles the hood of a sweatshirt, with the sheath forming first and surrounding the tip of the stem (Fig. 9.6). The opening of the hood reveals the younger leaves and apical meristem within, while the back of the hood projects upward as the beginning of the leaf blade. A second intercalary meristem often develops at the base of the young leaf sheath, enabling it to elongate into a cylindrical tube.

The shoot apex of a monocot thus consists of a series of concentric leaf sheaths that protect the apical meristem and young leaves in the center. All monocots, even massive palm trees, go through a juvenile phase in which the main bud remains below ground and only the sword-shaped blades of the leaves emerge. And even when the massive apical bud of a palm rises above

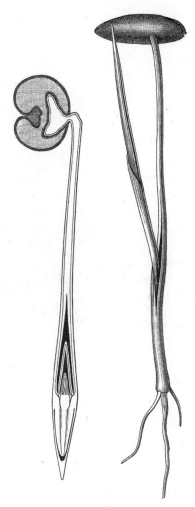

Figure 9.3 The monocot cotyledon usually stays within the seed and elongates at its base to push the embryo out into the soil. Modified from Kerner & Oliver 1895.

ground, the apical meristem remains deeply embedded within the many overlapping leaf sheaths.

With the soft new tissues forming at the base of the blade, the tip is the first part to harden, adapting it to push up through the soil as new leaf tissues are added below it. This growth is more-or-less indeterminate. It can be brief or extended, depending on conditions, and sometimes can reactivate if the upper part of the blade is destroyed. Since the leaves emerge from the soil through their own basal growth, the main stems and buds remain underground, insulated from heat, drought, and grazers by a thick layer of soil, as well as by the overlapping bases of the leaves themselves.

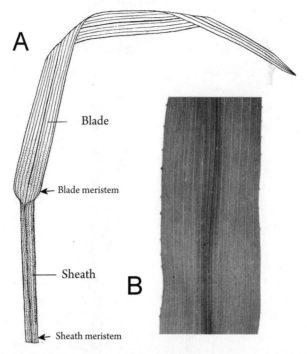

Figure 9.4 In the typical monocot leaf (A), a tubular leaf sheath expands into an elongate blade with parallel veins (B). Redrawn after Transeau et al. 1940.

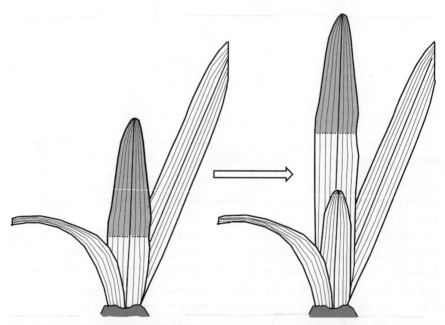

Figure 9.5 Like most monocot leaves, *Amaryllis* leaves grow from a basal intercalary meristem. A portion of a new leaf was marked (shaded area), and a week later, the marked portion had been pushed upwards by the new tissues produced at the base.

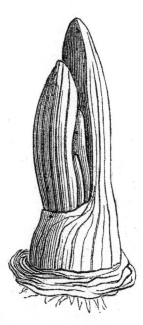

Figure 9.6 A very young monocot leaf consists of a prominent cylindrical sheath, with the rudimentary blade projecting from the back side. The next younger leaves emerge through the opening of the sheath. Both the blade and the sheath may elongate through a basal intercalary meristem. Drawing from Sachs 1874.

This contrasts with the leaves of almost all other (dicotyledonous) angiosperms, which typically have a narrow petiole and a broad blade. The petioles contain a relatively small number of vascular bundles that enter into the blade, and then repeatedly branch and reconnect to form a complex netted pattern as the blade expands. The smallest veinlets in a dicot leaf branch at right angles into the photosynthetic tissues of the areoles and end blindly (see Chapter 8).

Aside from the grasses, narrow, sword-shaped leaves are widespread among monocots, including cattails (*Typha*), daylilies (*Hemerocallis*), amaryllids (family Amaryllidaceae), orchids, irises (family Iridaceae), and countless others. More complex shapes can develop, however, particularly in specialized monocots where the stem rises above the ground. Palms, banana plants, and philodendrons, for example, have broad blades that may superficially resemble those in other angiosperms. These specialized leaves, however, represent modifications of the basic type, and their juvenile leaves are typically sword shaped.

Broader leaves can arise through expansion of tissues between the parallel veins, bowing the veins outward and strengthening the cross veins (Fig. 9.7). In some aroids and other vining monocots, like *Philodendron*, *Smilax* (catbriers), and *Dioscorea* (yams), the leaves may expand into heart shapes similar to those of morning glories and other dicot vines. The expansion of the vein system

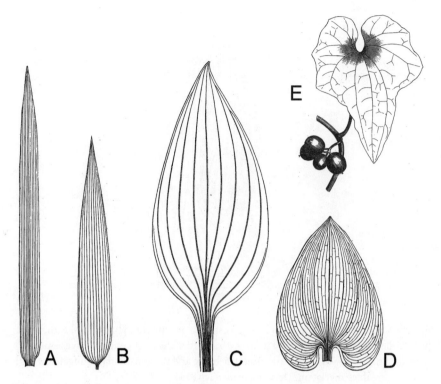

Figure 9.7 The common ancestor of modern monocots was apparently grasslike with parallel veins (A). As more specialized monocot leaves evolved, some became more like eudicots, with a portion of the blade becoming constricted into a petiole section (B–E), and sometimes with the upper blade expanding outward, making the parallel veins bow apart. Even when the latter happens, the veins enter the stem individually, and they remain independent of one another. Modified from Kerner & Oliver 1895 (A–D) and LeMaout & Decaisne 1876 (E).

sometimes results in branching at irregular angles, and a netted pattern of venation, strongly resembling that in dicot leaves (Fig. 9.8).

The leaves of the banana plant (*Musa* spp.), as well as of its relatives in the ginger (Zingiberaceae) and bird-of-paradise (Strelitziaceae) families, take on a broad, paddle shape, an adaptation for more light-gathering capability. The veins, however, are still parallel. Thousands of veins emerge from the underground rhizome and run parallel to one another through the cylindrical sheath. The veins converge to form a petiole, then diverge one by one at a sharp angle from the mass of veins in the central midrib to form the broad blade (Fig. 9.9). These broader leaves are adapted to tropical habitats where light gathering, rather than pushing up through the soil or regenerating missing tips, is the primary objective.

Giant tropical herbs like the banana are fundamentally like grasses, with an underground rhizome system and herbaceous upright shoots. Without

Figure 9.8 Among monocots, members of the aroid family (Araceae) have the most flexibility of leaf shape. The simplest leaves in the family, as well as the juvenile leaves throughout the family, are narrow and grasslike, as in typical monocots. In more complex leaves, the parallel vein pattern is distended, and the veins may branch in a netted pattern as in dicots. A. *Monstera deliciosa*, in which holes develop in the large leaves; B. a portion of a *Philodendron* leaf, showing a branching venation pattern. Drawing from Thomé 1877.

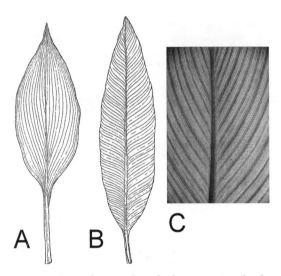

Figure 9.9 Some monocot leaves become broader by expansion that bows the parallel veins outward (A), with some of the veins extending up the middle before diverging; in the banana leaf and its relatives (B, C), the veins remain part of a strong midrib until diverging, one by one, to the sides. Drawings from Ganong 1916.

drought, fire, or grazers to contend with, however, the upright shoots may live for several years. Banana plants may be five meters or more in height, but they are completely herbaceous. Cutting horizontally through a banana "trunk" reveals rings very much like those in an onion bulb (Fig. 9.10). Each ring of this pseudostem represents a long cylindrical leaf sheath, which is topped by a spreading leaf blade. The plant quickly becomes treelike as each sheath pushing up through the center grows a little taller than the previous one. This can happen in a year or less. At maturity, a flower stalk pushes up through the center.

Even more complex patterns develop in a palm leaf, which is essentially "corrugated" for strength. As in the banana leaf, a mass of parallel veins runs through a long petiole and into the midrib. The veins then diverge to the sides, in parallel groups that constitute folded units. In a few understory palms, this results in a large, undivided leaf—essentially a corrugated banana leaf. But in most palms the leaflets are separated, creating a truly compound leaf, which is rare among monocots. A compound leaf provides less resistance to the wind and also is less likely to overheat in full sunlight.

The corrugation appears as ripples along each side of the young leaf primordium (Fig. 9.11A). In palms with pinnate ("feather-like") leaves, the midrib elongates between the folds and separates them into distinct leaflets (Fig. 9.11B, C).

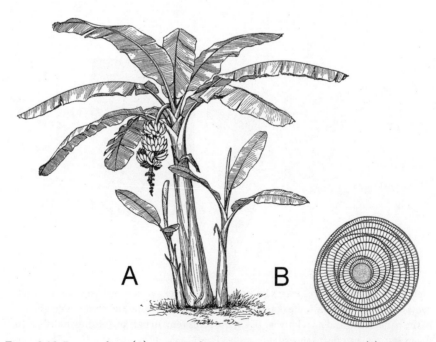

Figure 9.10 Banana plants (A) are tropical evergreen perennials. The "trunk" (B) of the upright shoot is a false stem made up of overlapping leaf sheaths. Drawings from Brown 1935.

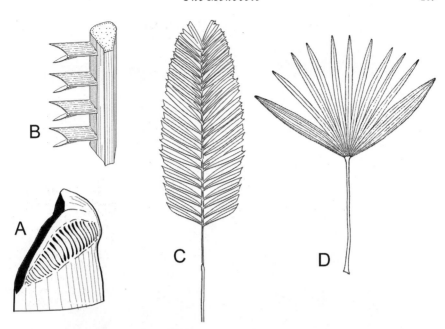

Figure 9.11 The complex palm leaf begins as a simple, hood-shaped structure similar to grasses and other monocots, but wrinkles develop along the sides (A) that deepen into accordion-like folds. As the leaf lengthens and matures, the folds separate to form a pinnate (B, C) or palmate (D) compound leaf. Redrawn after Corner 1966 (A–C) and Ganong 1916 (D).

In others, the midrib does not elongate and the folded leaflets fan out from the petiole tip (Fig. 9.11D).

The monocot stem

Typical monocot stems contain numerous vascular bundles that appear to be randomly scattered throughout the interior, while dicots typically have a small number of bundles arranged in a distinct ring (See Chapter 8, Fig. 8.3). The many parallel bundles of each monocot leaf emerge at the same time from the stem after having arched through the center of the stem from their attachment to bundles further down. This pattern is the result of the way monocot stems increase in thickness. The arching of the bundles is created just below the broad tip of the stem in a meristematic zone referred to as the primary thickening meristem. The stem achieves its full thickness, up to a meter in some palms, in this zone as cells here divide and expand horizontally. Each new ring of bundles enters the zone from a small embryonic leaf in the center of the broad shoot apex, and is gradually displaced outward as the base of the leaf expands (Fig. 9.12A–C). Overlapping sets of arching bundles from different leaves create the appearance of random distribution in the stem cross-section (Fig. 9.12D).

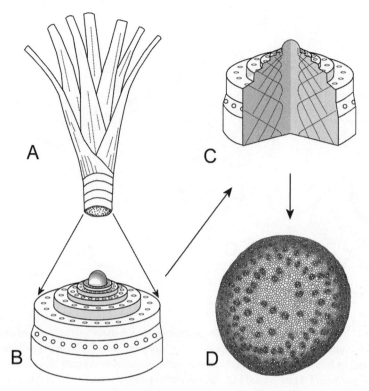

Figure 9.12 The leaf sheaths of monocots typically wrap around the stem (A), attaching to the top of the stem through concentric rings of vascular bundles (B); within the stem tip (C), vascular bundles enter at the center of the stem, connect to an older bundle lower in the stem, and are bent outward as the stem tip expands. This results in the scattered appearance of bundles (D). Drawing (D) from Brown 1935.

Geophytes

The fireproof, grazer-proof grasses have underground rhizomes that can go into a dormant state under adverse conditions. Many monocots go dormant on a predictable seasonal schedule, and the exposed parts of the plants virtually disappear during the off season. Perennial herbs of this nature are called geophytes ("earth plants"). Many monocots adapted for periodic dormancy, such as the bearded iris or common ginger (Fig. 9.13), have thick rhizomes filled with stored food reserves that allow for rapid growth or reproduction at the beginning of the next season.

In other geophytes, more compact underground organs have evolved. A corm is a short, roundish underground stem, a modified rhizome with its apical bud(s) facing upward. It consists of solid stem tissues filled with stored nutrients, and may be roughly spherical or quite flattened (Fig. 9.14). Corms are particularly common in the iris family (Iridaceae), and the ornamental *Gladiolus* is a common

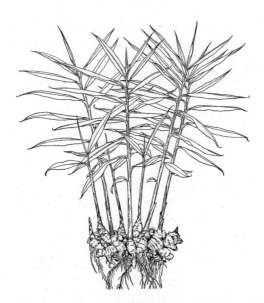

Figure 9.13 Ginger rhizome segments are short and thick, and branch to form dense clumps. Leafy shoots are produced each season then discarded. Drawing from Brown 1935.

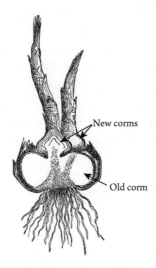

Figure 9.14 Corms of crocus and other members of the iris family are dense, food-filled, underground stems, with adventitious roots emerging from their lower surfaces. New plants form on top of the old corms each year, and the older corms gradually disintegrate. Drawing from Ganong 1916, attributed to Figurier.

example. Corms typically form in vertical succession, a new corm forming on top of an old corm at the beginning of each season. Specialized contractile roots then pull the new corm downward, filling the void as the old corm disintegrates. Corms can multiply into an extensive clump through the formation of small cormlets on their sides.

Bulbs appear similar to corms on the outside, but consist of densely packed modified leaves rather than solid stem tissues. The actual stem of the bulb is just a small disk at the base. In the true lilies (genus *Lilium*), a rather loose bulb forms from short, specialized storage leaves at the base of the shoot (Fig. 9.15). In *Amaryllis* or onions, however, the bulb forms from the circular, tightly nested sheaths of ordinary leaves, which swell with food reserves over the growing season. As the leaf blades die off, the thick sheaths remain as the overwintering bulb. An "onion ring" is a cross section of one circular, food-filled leaf sheath.

Epiphytes

The clonal growth form of monocots has proven adaptable to the trunks and spreading branches of tropical trees. Plants that live on other plants are called epiphytes. Three families of monocots have been particularly prolific in this subhabitat: the orchids (Orchidaceae), bromeliads (Bromeliaceae), and the aroids (e.g., *Philodendron* and *Anthurium*, Araceae). There are even a few gingers (Zingiberaceae) who have taken to arboreal life. Epiphytic monocots are confined to tropical and subtropical forests, while temperate rain forests more commonly host epiphytic ferns and bryophytes.

The branches of rainforest trees provide a rather harsh environment for plants, however. Though it may rain every day in equatorial regions, intense

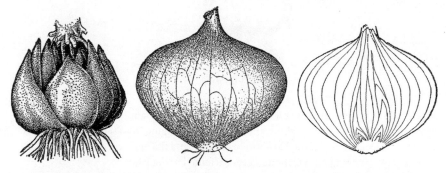

Figure 9.15 Bulbs are compact underground clusters of leaves or leaf bases modified for storage of food and water. In lilies (left), bulbs consist of short, specialized leaves modified for underground food storage, while in members of the onion and amaryllis families (middle and right), the bulb is made up of the swollen bases of ordinary photosynthetic leaves. Drawing from Brown 1935.

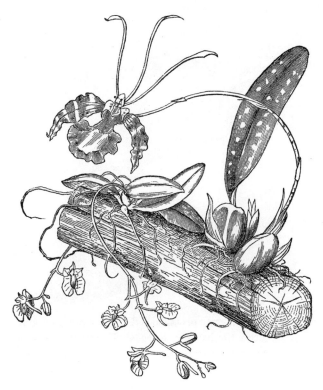

Figure 9.16 Epiphytic orchids typically have thick, succulent leaves and/or swollen bases called pseudobulbs, which are adaptations for storing water. Drawing from Gray 1879.

sunlight and warm temperatures can dry out the epiphytes' habitat quickly during the day. Orchids adapt to this challenge with thick, water-filled leaves, sometimes including thickened "pseudobulbs" at the base of the petioles, as well as thick, water-storing roots (Fig. 9.16). Aroids also have thick roots and stems that store water. Bromeliads, on the other hand, have thin, flat leaves that fit tightly together to form a water-holding cup (Fig. 9.17). Adventitious roots grow upward into the cups to absorb water. The so-called Spanish moss is a bromeliad that has neither cups nor roots, but the entire plant absorbs water during rainy times, then desiccates during dry times, mimicking the lifestyle of true mosses (Chapter 3).

Monocot trees

The ability to form wood is completely absent among monocots, and so exceptionally large monocots like palms and bamboos have evolved new ways to support their upright growth. Though many larger palms (family Arecaceae) have single trunks, a more general pattern in the family is multiple upright stems produced over time from an underground rhizome system (Fig. 9.18A) as in other

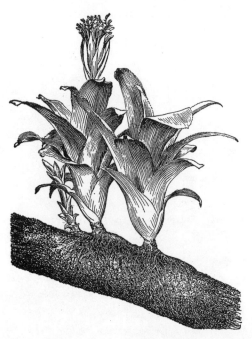

Figure 9.17 Bromeliads are evergreen perennial herbs of the pineapple family (Bromeliaceae). Many, such as this *Billbergia*, are epiphytic; their rhizomes creep along the branches of trees, and their roots serve primarily for attachment. Drawing from Brown 1935.

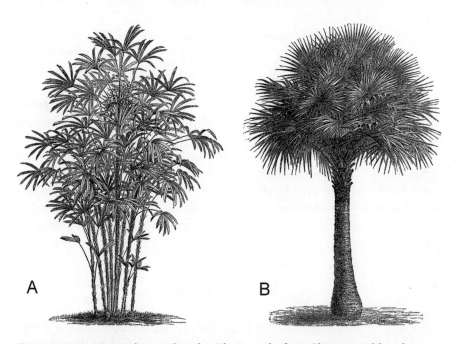

Figure 9.18 A. Many palms, such as this *Rhapis excelsa* from China, grow like other monocots from underground rhizomes. Their upright shoots, however, are long lived and fibrous. B. Palms such as *Livistona australis* focus all their growth into a single upright shoot and are no longer capable of sprouting new stems from the base. Drawings from LeMaout & Decaisne (A) and Thomé 1877 (B).

monocots. *Nypa* palms, which form colonies in brackish mangrove swamps in southeast Asia, and saw palmetto (*Serenoa repens*), which forms thickets throughout Florida, represent a slightly different growth form, one that might represent the form of the original palms: thick, horizontal, branching stems turning upward at the tips to spread their rosette of leaves. Palms like a coconut or date palm, have abandoned their rhizome systems, focusing growth in a single trunk (Fig. 9.18B), but they do it without wood or an axial root system. They are supported by thick, long-lived adventitious roots at the base of the trunk.

Arborescent palms become treelike by concentrating growth in a single large bud, which results in an exceptionally broad primary thickening meristem (see Fig. 9.12), rather than by forming layers of wood over time. They are thus more or less the same diameter from top to bottom. The stems are strengthened with dense bundles of fibers, which support their upright growth and crown of large, heavy leaves. This seems to be fully as effective as any woody trunk, and often, it is more resilient. With their cable-like supporting fibers, palms can sometimes bend to the ground in a strong wind and right themselves undamaged. Members of the tropical screwpine family (Pandanaceae) become trees with the support of extraordinarily massive adventitious prop roots (Fig.9.19). Interestingly, their leaves have remained long and strap-shaped, not becoming compound like palms. This is balanced by the fact that screwpines generally branch to form more bushy crowns. Their stems are limited by the lack of a cambium, so as they branch, they become progressively narrower.

Bamboos evolved the stature of trees quite differently, with a light-weight design featuring hollow stems reinforced with extremely dense bundles of fibers. Such stems have been used by indigenous cultures for a vast array of structural purposes, from house construction to chopsticks. All bamboos form colonies via horizontal rhizomes. Their success as trees is largely due to their rapid elongation through simultaneous basal growth of many internodes (Fig. 9.20). This is a growth form that first appeared among ancient giant horsetails (see Chapter 4).

A few monocots, like the dragon tree (*Dracaena draco*) and other members of the *Dracaena* and aloe families, have become trees by evolving a new form of secondary thickening—not one that adds layers of wood, but rather one in which whole new vascular bundles containing xylem, phloem, and fibers form continually around the periphery of the stem. These trunks get wider with age, like dicot or gymnosperm trees, and can branch to form broad leafy crowns (Fig. 9.21).

Monocot trees also hold one of the many types of longevity records. Woody trees like bristlecone pines may live as individuals for thousands of years, and seagrasses form clonal colonies that may be up to 200,000 years old, but both do so by continually creating new tissues and organs. A woody tree continually produces fresh phloem and xylem tissue from its vascular cambium, and the seagrass produces fresh rhizome segments as older ones decay. As a palm tree gets taller, however, no new tissues are produced in the older parts of the trunk. Living phloem tissues,

Figure 9.19 A. Members of the Pandanaceae are treelike monocots without secondary growth. Large adventitious prop roots provide support and transport of water and minerals from the soil; this would be accomplished through secondary growth in true woody plants. B. The massive prop roots of a *Pandanus* in Papua New Guinea dwarf a human being. Drawing from Brown 1935.

which carry carbohydrate and other substances from the leaves to the roots, function for the life of the palm, several hundred years in some cases (Tomlinson et al. 2012), and therefore contain the oldest known living cells among plants.

Xerophytic Monocots

Savannas, which are home to vast stands of grasses, are harsh environments that may be bone-dry for half the year. The even drier habitats we call deserts support grasses more fleetingly, if at all. Deserts tend to be dominated by hard-leaved shrubs and succulent cacti, but monocots have made a presence here also. Xeric vegetation is too sparse to support fire or large grazing mammals, so the challenge for monocots shifts from underground survival and regeneration to water conservation. This has resulted in evergreen succulents like *Aloe, Agave, Yucca, Sansevieria,* and *Xanthorrhea* (Fig. 9.22).

Figure 9.20 In bamboos, hollow, rapidly elongating upright shoots arise from underground rhizomes. These shoots, though nonwoody, are heavily fibrous, and live for several years. Drawing from Thomé 1877.

Water conservation is enhanced in many of these plants by crassulacean acid metabolism (CAM). In this process, stomata are open for gas exchange during the night, greatly reducing the loss of water through evaporation. Carbon dioxide is incorporated into 4-carbon molecules of malate, which are stored until daylight. They are then released and used to synthesize carbohydrate.

In warm climates, many grasses have adopted a similar process, usually referred to as "C_4 photosynthesis," that helps them conserve the carbohydrates they produce and allows them to be much more productive. Carbohydrates produced through photosynthesis are often wasted by a process called photorespiration, caused by the presence of oxygen also generated by photosynthesis—they are burned up on the spot with no apparent benefit. In the C_4 process specialized cells that only produce carbohydrate are organized in rings surrounding the vascular bundles. They are thus isolated from cells in the rest of the leaf that are engaged in the light reactions and releasing oxygen. A temporary 4-carbon molecule similar to that in CAM shuttles carbon dioxide into the carbohydrate-producing cells.

Figure 9.21 The dragon tree, *Dracaena draco*, is a monocot with a novel form of secondary growth, in which a cambium-like layer produces whole vascular bundles embedded in parenchyma tissue. The species is native to the Canary Islands and other islands off the coast of Africa. Drawing from Brown 1935, attributed to Bailloon.

Aquatic Monocots

We began our survey of monocots with grasses growing in the savannas, and conclude with those adapted to aquatic habitats. Theorists over the years have generally believed that monocots began either in the aquatic environment or in a drier savanna. The truth may lie somewhere in-between, and we'll address that question after a quick look at the great variety of aquatic monocots.

Plants exhibit a wide range of tolerance or requirement for wet conditions, with some fully submerged and others living at the edges of ponds or in seasonally wet marshes. We can define aquatics broadly to include plants in which the roots are in water-saturated soil at least part of the time,[1] and monocots have excelled in every conceivable aquatic lifestyle. Roots and rhizomes need oxygen for basic metabolic functions, but soil that is saturated with water is generally oxygen-deprived. Aquatic monocots (as well as the unrelated water lilies) have therefore evolved systems of air channels (aerenchyma) that carry oxygen from the leaves to the roots (Fig. 9.23).

The majority of aquatic monocots are in the order Alismatales, to which the sea grasses, arrowheads (*Sagittaria*), and more terrestrial aroids (Araceae) belong.

Figure 9.22 The genus *Aloe*, along with *Agave, Yucca*, and other monocots adapted to deserts, often have succulent leaves adapted for water storage and conservation.

Aquatics have evolved independently in the unrelated Pontederiaceae (pickerelweed, water hyacinth), Juncaceae (rushes), Typhaceae (cattails), Cyperaceae (sedges, papyrus), and even some grasses (Poaceae), as well. Some are like water lilies, with long petioles and roundish blades that rise to the surface from underwater stems (e.g., *Hydrocharis*; Fig. 9.24). Some are fully submerged in fresh or even salt water, with sword-shaped leaves arising from strong rhizome system ("sea grasses" like *Vallisneria*; Fig. 9.25). Other submerged plants, like *Elodea* or *Hydrilla*, bear short, sword-shaped leaves on elongate stems (Fig. 9.26). Marginally aquatic species grow around the edges of the water and typically have broader, often arrowhead-shaped, blades (Fig. 9.27A). They are found mainly in

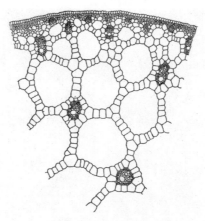

Figure 9.23 Aerenchyma tissue consists of parenchyma tissue in thin sheets lining large air spaces or canals. It is common in aquatic plants, serving both for buoyancy and conducting air down to the roots. This example is from the inflated lower petiole of the water hyacinth. Drawing from Brown 1935.

Figure 9.24 *Hydrocharis* has roundish, floating leaves like those of water lilies, though the plants are not rooted in the mud. The leaf shape develops through dilation of the basic monocot pattern of parallel venation. Drawing from Kerner & Oliver 1895.

Figure 9.25 Sea grasses have grasslike leaves arising from rhizomes rooted in the mud. Pictured is a species of the freshwater genus *Vallisneria*. Drawing from Kerner & Oliver 1895.

Figure 9.26 *Elodea* has elongate stems with narrow, whorled leaves, superficially resembling *Cabomba* or *Ceratophyllum*. Photograph by Christian Fischer, posted on Wikimedia Commons, licensed by Creative Commons.

the Alismataceae, Araceae, and Pontederiaceae. Still others float on the surface, with roots dangling in the water. These include the duckweeds (*Lemna*), water lettuce (*Pistia*), and the notorious water hyacinth (*Eichhornia*) (Fig. 9.27B).

In papyrus (*Cyperus papyrus*), a modified flower stalk (peduncle) lifts a tassel of photosynthetic branches above the fluctuating water level. The stalk rises three meters or more through basal growth of a single internode (Fig. 9.28). This long stalk, uninterrupted by nodes, is run through by numerous straight bundles of fibers, which were exploited by the ancient Egyptians for making paper, basketry, and even boats (not to mention floating bassinets for future prophets!). The history of the western world may have been radically different without this unique plant.

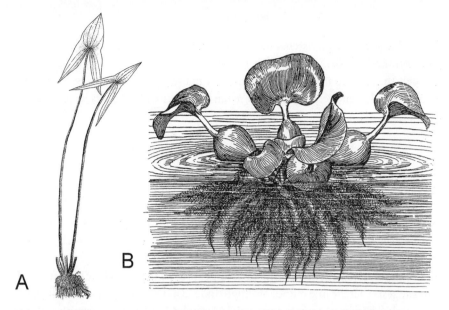

Figure 9.27 Aquatic monocots often have emergent leaves with expanded blades. *Sagittaria* (Alismataceae) (A) leaves expand into an arrowhead shape. The water hyacinth (*Eichhornia crassipes*, Pontederiaceae) (B) is a floating aquatic plant that has become quite a nuisance in some places. The lower parts of its leaf petioles are expanded and filled with aerenchyma tissue, which gives them buoyancy. Redrawn after Shaukat in Flora of Pakistan http://www.efloras.org, Missouri Botanical Garden, St. Louis, MO (A), and Brown 1935 (B).

Figure 9.28 Tassels of photosynthetic branches are raised into the air by the elongation of a single internode on each upright shoot of the Egyptian papyrus, *Cyperus papyrus*. Drawing from Kerner & Oliver 1895.

Basal intercalary meristems have thus proven to be as valuable for elevating photosynthetic tissues up through the water into the air as for pushing them through the soil. So it is not surprising that many monocots have become marginally or fully aquatic. But does that mean that the first monocots were aquatic?

Where did the Monocots come from?

In recent phylogenetic studies (Davis et al. 2004, Graham et al. 2006), the marginally aquatic genus *Acorus* appears to represent the most ancient lineage of monocots. So it is in the same position ("sister group") to the rest of the monocots as *Amborella* is to angiosperms as a whole. Once again we must ask which features of *Acorus* are archaic and which are specialized.

The largely aquatic order Alismatales appears to be the next clade to branch off. This suggests that the first monocots might indeed have been aquatic. Phylogenetic studies also, however, point to terrestrial Magnolids, not aquatic Nymphaealeans or any aquatic eudicots, as the closest relatives of the monocots (see Stevens 2001 onwards). So ancestral monocots most likely evolved from terrestrial plants that had broad leaves with branching veins.

We can be sure that monocots did not evolve directly from either water lilies or submerged plants like *Cabomba* or *Ceratophyllum*, not only because of the DNA-based phylogenies, but also through a simple comparison of leaf structure. Some aquatic monocots do have broad leaves like very much like water lilies (as in *Hydrocharis*, see Fig. 9.24), but these develop through the distortion of a basic parallel-veined pattern, like other monocots discussed earlier. The same is true of the arrowhead-shaped leaves of some species of *Sagittaria* (Alismataceae), which have a long petiole in which the many parallel veins are crowded together before diverging toward the three corners of the leaf (see Fig. 9.27A).

Monocots with submerged, leafy stems, like *Hydrilla* or *Elodea*, have short, sword-shaped leaves with parallel veins (see Fig. 9.26), contrasting with the branched and dissected leaves of *Cabomba* (Nymphealeaes) or other submerged dicots (see Chapter 8, Fig. 8.23). Thus, aquatic monocots evolved habits similar to those of their distant dicotyledonous cousins through convergent evolution and are not directly related to any of them.

As far as we can see, the common ancestors of all known monocots had sword-shaped leaves with parallel veins. These ancient monocots, living some 120–140 million years ago, were most certainly clonal plants with underground rhizomes lacking in secondary growth and with wholly adventitious root systems. They were more like cattails or sea grasses than *Hydrocharis* or *Elodea*.

Among the nearest living relatives of monocots, the magnolids, there are clonal perennials with rhizomes, such as *Asarum* ("wild ginger," Aristolochiaceae) and *Peperomia* (Piperaceae), and some even have flowers with parts in threes, like most monocots. These give us clues as to what might have preceded the monocots. No magnolid has sword-shaped leaves with parallel veins, however, so the transition to that new form of leaf (and cotyledon) was the key event in the origin of the monocots as we know them. What then were the selective forces that favored that transition?

Evolution of the first sword-shaped blade most likely involved the gradual loss of the original broad blade, confinement of growth to the base, and widening of the petiole to become a new form of blade. This shift was most likely driven by frequent loss of the original blade to fire, grazing, and other hazards.

Whether the primary shift of growth to a basal intercalary meristem happened first in the single cotyledon or in the juvenile leaves that emerge from the ground, is unknown. Both are adaptive to the underground clonal habitat and must have happened in close conjunction. Stebbins (1974) emphasized the fact that the monocot cotyledon usually remains within the seed, elongating at the base to push the rest of the embryo out of the seed and often deeper into the soil (see Fig. 9.7), rather than to push itself above ground. This underground maneuver is possible when the soil is moist and soft, but the adaptive value comes later when the soil dries out, as the terminal bud and embryonic rhizome have been "planted" at a more secure depth. This suggests a habitat in which the soil alternates between wet and dry—conditions we would find in a savanna or seasonal marsh, not in a permanent aquatic habitat.

The hard-tipped, sword-shaped leaf pushing up from the underground stem also would seem to be highly adaptive in a seasonally dry habitat but not in a submerged aquatic environment. The shift from broad blades to narrow blades growing from the base makes most sense as a response to the grazing and fire found in the savanna. Aquatics don't face the same pressures and can take advantage of the greater light-gathering capacity of broader leaf blades. Broad blades, in fact, have evolved secondarily numerous times in aquatic monocots through distortion of the basic parallel pattern (*Sagittaria, Hydrocharis, Pontederia*, etc.). By implication then, there would have been no reason for a radical change in leaf shape if monocots had evolved in the water directly from aquatic dicot ancestors. Petioles that elongated through intercalary meristems would have sufficed.

The fossil record so far does not help us much with this question, but can we learn anything from plants alive today that are adapting to similar conditions? An unexpected example comes from the eudicot carrot family (Araliaceae). Most members of the family have relatively broad leaf sheaths and petioles with many parallel bundles, as in monocots, but they also have rather complex leaf blades with branching veins like other eudicots. A celery stalk, which is in this family, is a good example.

In *Eryngium yuccifolium*, however, the compound leaf blade is suppressed, and the leaf sheath is flattened and lengthens through basal growth into a monocot-like blade with parallel veins (Fig. 9.29). *Eryngium* grows in prairies

Figure 9.29 *Eryngium yuccifolium* is a eudicot in the carrot family that appears to be retracing the early evolution of the monocot leaf.

that are sometimes flooded, and so perhaps represents a re-creation of the of the original monocot leaf type. Marginal aquatic habitats are sometimes dry, while savannas may at times be flooded. The example of *Eryngium yuccifolia* supports the hypothesis that the monocots evolved in a habitat somewhere between savanna and wetland. Interestingly, the seedling of *E. yuccifolia* has two ordinary cotyledons, like other dicots, but the first foliage leaves are grasslike (Fig. 9.30), suggesting that the monocot leaf may have evolved its distinctive growth form before the single cotyledon did.

So if early monocots had sword-shaped leaves and lived in savannas, does it mean that they were in fact grasses? "Protograsses" might be a better term. They were like grasses in their general architecture and the shape of their leaves, but modern grasses have developed a number of additional adaptations that contribute to their dominance of the savannas, including wind pollination.

So back to the question of what is archaic about *Acorus*. How much does it resemble the first monocots? *Acorus* (Fig. 9.30) and the Alismatales can be considered the "ANITA grade" of monocots. Two members of the Alismatales in particular, *Tofieldia* and *Butomus* (Fig. 9.31), show some similarities with *Acorus* and may provide some further clues about the most ancient monocots. *Acorus* and *Butomus* live at the edge of the water, with their rhizomes and roots usually in

Figure 9.30 The *Eryngium yuccifolium* seedling has two cotyledons, but a very grasslike shoot. Courtesy of www.grownative.org, a program of the Missouri Prairie Foundation.

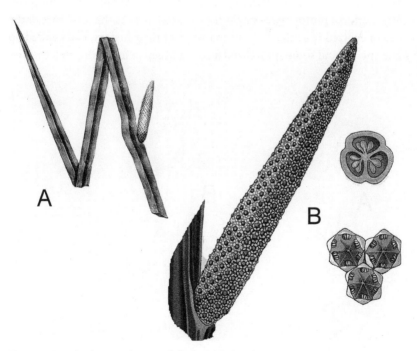

Figure 9.31 The leaves of *Acorus* (A) are folded and sealed together, with the flowering stem emerging from the sheath. The flowers are packed onto a thick spike (B), and the carpels are completely fused into a solid pistil. Drawings from Barton 1818 (A) and Thomé 1885 (B).

water-saturated mud, while *Tofieldia* grows in meadows that are sometimes wet and sometimes dry. All have simple, bisexual, insect-pollinated flowers. They all also have long, sword-shaped leaves, but with an interesting twist: the blades are folded and the two halves are fused together above the sheath, forming a unifacial (single-faced) leaf—that is, both sides of the leaf above the sheath are actually the back side. This presumably is a specialized feature, but one that evolved very early. It later evolved again in the Iris family (Iridaceae), but most monocots, including other members of the Alismatales and the grasses, have ordinary unfolded and unfused (bifacial) leaves.

Tofieldia and *Butomus* are also presumably unspecialized in that their carpels are more-or-less free from one another (apocarpous) [though this has been contested[2]], as they are in many of the Alismatales and archaic angiosperms in general, while the carpels of an *Acorus* flower are solidly fused together (syncarpous) (Fig. 9.32). *Tofieldia* also has more primitive vessel members, with slanted endwalls and ladderlike crossbars. The flowers of *Acorus* are densely packed into a thick spike, another specialized feature seen elsewhere in the Araceae, Pandanaceae, Cyclanthaceae, and some palms. So *Acorus*, the lone surviving descendant of the oldest monocot clade, is more specialized in several ways than are some members of the Alismatales. But together, *Acorus*, *Butomus*, and *Tofieldia* give us a glimpse of what early monocot protograsses might have looked like. As the forebears of the three genera evolved over the 100 million years of their history, each ended up with some ancient and some specialized features. Keep in mind also that the first

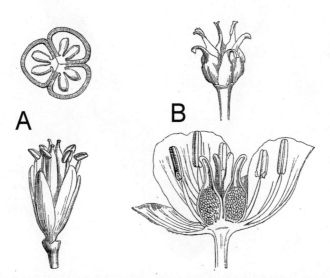

Figure 9.32 *Tofieldia* (A) and *Butomus* (B) flowers have 3 separate carpels. Those in *Tofieldia* are pressed together more tightly, but not fused into a solid pistil. The carpels of *Butomus* are specialized in having numerous small ovules distributed throughout the inner walls. Drawings from LeMaout & Decaisne 1876.

actual monocots would have been "stem monocots" that preceded the common ancestor of these genera.

Other forms of monocots evolved as protograsses migrated into wetter, drier, colder, and tropical habitats. Clearly they did so quickly into submerged aquatic environments, but probably equally quickly into drier habitats. The palm family also appears to be quite ancient, and some palms also have separate carpels (apocarpy), including the mangrove palm of the old world tropics, *Nypa fruticans*. Actual fossil evidence of the largely herbaceous monocots is rather scanty, however. The origin of the grass family can be dated to about 60–55 million years ago because the ample production of wind-dispersed pollen by grasses has left an abundant microfossil record (Kellogg 2001), but there is little record of insect-pollinated monocots.

So we have ended this journey with a group of plants that have come to dominate many habitats of the Earth through a unique and specialized, but highly versatile, architectural plan. They are angiosperms with advanced anatomical and reproductive structures, but their superiority as clonal plants in stressful and extreme environments points to them as a major cutting edge in plant evolution. If the world survives the current plague of *Homo sapiens*, it is likely that the diversity and dominance of monocots in habitats ranging from marine meadows to deserts and tropical forests will continue to grow.

EPILOGUE

We have seen in nine brief chapters how the distinctive features of plant life have evolved sequentially as various environmental challenges and opportunities were met. Photosynthesis evolved as a means to gather previously untapped energy resources in a world where such resources were limited. Nitrogen fixation followed as a means to expand the extremely limited availability of this element to life. The spectacular process of endosymbiosis allowed larger organisms to come into existence, eventually becoming multicellular and moving onto land.

Early land plants succeeded in their conquest by evolving protection from desiccation, new kinds of spores and spore-producing bodies, and internal chambers for sexual reproduction and embryo development. Independent sporophytes with vascular tissues evolved ever larger growth forms in response to competition for light and spore-launching advantage, creating the first forests. Seeds and then flowers evolved in response to the challenge of reproducing and dispersing one's offspring in the terrestrial environment while remaining fixed to one spot of ground.

In the final three chapters, we saw how flowering plants further condensed their reproductive cycle, enabling them to evolve into a wide variety of growth forms, including short-lived herbs, aquatic plants, epiphytes, and xerophytes, as well as more efficient trees. We saw how they also took advantage of the animals that would feed upon them to further expand their means of pollination and dispersal. Families such as the sunflowers, legumes, and grasses appear, for differing reasons, to be on the cutting edge of plant evolution and to be still rapidly evolving new forms.

The question then arises, "what is the future of plant evolution?" The answer is heavily tied to the fate of our own species. Pundits, politicians, scientists, and fiction writers have given us an abundance of futures to fear or choose from. In the best-case scenario, we stabilize our numbers, halt the degradation of air, water, soil, and climate, and enter into a civilization based on sustainable economics and renewable resources. We vigilantly protect the remaining natural ecosystems and allow the natural course of evolution to proceed. Grasses might continue to expand their dominance into new habitats. Legumes might continue to diversify and integrate into more ecosystems where nitrogen-poor soil limits productivity.

Other plants and animals might respond to the opportunities created by these dominant trendsetters, and the coevolutionary tit for tat would continue. Certainly the kaleidoscope of flower forms and animal pollinators would continue to shift and diversify, as would defensive plant chemicals and animals adapted to tolerate them. Dark-horse members of the plant kingdom might emerge and dramatically introduce new life strategies unheard of today. Unfortunately, we would not see it happening, for the adaptive results of the evolutionary process emerge too slowly for us to perceive. Only if our descendants survive for another 10–20 million years might they be able to document the emergence of new forms of plant life by comparing then with now.

Alas, most of us do not see such a future, perhaps not even a future that includes humans for very long. If we do continue for a few more centuries, bumbling along focused on short-term interests, the noose will continue to tighten on the natural world. More cities, parking lots, athletic fields, and farmland means less space for natural communities, which will dwindle to the point of nonsustainability and mass extinction. The continued evolution of plants will take two forms. Cultivated crops will be increasingly "improved" through both traditional breeding and genetic engineering, but the underground society of plants—weeds and invasive exotics—will also continue to expand and adapt to the margins of our controlled habitat. Genetics may allow us to know and preserve the code for what biological diversity remains. DNA in a bottle does not evolve any more, but it might be used to resurrect species, here or on another planet.

When we do manage to exterminate ourselves, the pampered agricultural crops will most likely fade away, and the weeds will rise to dominate. The future of plant evolution in this scenario will belong to crab grass, dollar weed, punk trees, and dandelions. Their descendants will recolonize the world, covering over our rubble, and will eventually diversify once again into a rich and varied flora. It has happened before.

Even that scenario is too optimistic for some. If human existence ends with an environmental catastrophe, much of the rest of the world's life will go with it. We may be so thorough in our departure that virtually all forms of life become extinct. But if anything survives—the hardy spores of some anaerobic bacterium perhaps—it could start over, reinvent photosynthesis, and eventually repopulate the Earth with a diversity of organisms. There is some speculation that life first arrived on Earth in the form of such spores, blasted perhaps from the surface of Mars by giant asteroids, surviving even the vacuum, radiation, and temperature extremes of space. So conceivably, some bacteria could survive a nuclear holocaust or the death of our oceans and atmosphere. How similar the diversity that would arise from that new beginning would be to what we have today is a wide-open question. Humanlike creatures might ultimately evolve again and have another chance to coexist with the natural world.

One way or another, the future of plant evolution is in our hands.

NOTES

Chapter 1

1. See Willis and McElwain 2002 for a discussion of stromatolite formation and history; there is also a website devoted to Shark Bay in Western Australia (www.sharkbay.org) that includes facts, photos and a video swim through a grove of stromatolites.
2. Oxidation, and its opposite, reduction, are fundamental processes in chemistry. You know that a shiny new pruning saw carelessly left outside under a hedge will slowly corrode, turning into iron oxide, or rust. This is one form of oxidation. Burning wood is another more dramatic form, as are the metabolic breakdown of food for energy by animals, food spoilage by microorganisms, and vegetable oil becoming rancid. Conversely, sugar (and ultimately wood, food, and vegetable oil) is formed through the reduction of carbon dioxide during photosynthesis. Energy is released from organic molecules during oxidation (e.g., as heat), and stored in organic molecules through reduction. Oxidized substances (e.g., carbon dioxide) are thus low-energy materials and less "useful" than their reduced counterparts like carbohydrates.
3. Technically, chemosynthetic organisms are called chemoautotrophs, lithoautotrophs, and other variations, while photosynthesizers are called photoautotrophs. "Autotroph" refers to the ability to make sugar from inorganic carbon dioxide, "photo" denotes light as an energy source, while "chemo" denotes a chemical source of energy, usually implying inorganic chemicals. "Litho" refers specifically to rocks or inorganic minerals and so is less ambiguous.
4. Archaea constitute another major group of prokaryotes that split off early from the "true" bacteria. It is most likely that higher forms of life, eukaryotes, evolved from archaean ancestors (See Chapter 2).

Chapter 6

1. Some authors prefer to shorten this to "ANA" as an acronym for the names of the three member clades: Amborellales, Nympheales, and Austrobaileyales (see Frohlich and Chase 2007). Many, however, prefer to continue using the ANITA acronym, even though the "I" (Illiciales) and "T" (Trimeniaceae) members are no longer recognized as distinct clades, but rather subgroups of the Austrobaileyales. The bulk of the angiosperms evolved after these basal clades and fall into three much larger groups: the Magnoliales, Monocots, and Eudicots. The first "A" in either acronym stands for the lowermost clade of the angiosperm tree, represented by a lone surviving species, *Amborella trichopoda*.

2. Drupes contain a single seed surrounded by a stony layer commonly referred to as a pit. Berries, by definition, have seeds embedded directly into the soft tissues. Both kinds of fruits are adapted to be eaten by animals, which either pass the seeds through their digestive tract or regurgitate them if they are too large to pass through.

Chapter 8

1. Under the rules of modern phylogenetic taxonomy, formal taxonomic categories consist of *a common ancestor and all of its descendants*. Such categories, or taxa, are by definition monophyletic. Though it sounds simple and obvious, this new definition caused a firestorm in the biological community, because it required a great deal of reorganization of our system of classification. Previously, organisms had been grouped together by overall similarity or key characteristics, and these didn't always reflect the pattern of evolutionary, or phylogenetic, history.

    With the greatly improved information about that pattern that emerged in the latter part of the 20th century, the dicotyledons proved to be a complex group that no longer fit the definition of a monophyletic taxon. The distinctive characteristics of dicots are actually characteristics of several groups of angiosperms, including the eudicots, magnolids, and the ancient ANITA grade. Their common ancestor is in fact the same as the common ancestor of all angiosperms, including the monocots. So by separating out the monocots, the Dicotyledonae does not contain all of the descendants of its common ancestor and therefore is not monophyletic, but rather paraphyletic.

Chapter 9

1. This is essentially the defining feature of "wetlands," which generally are under protection by law. Wetlands are vital to the water cycle and are therefore important for maintaining water supplies for human populations and agriculture. They also prevent erosion, especially along coasts, and provide vital habitat and breeding grounds for unique forms of wildlife and commercial fisheries alike. Therefore, identification of wetland or aquatic plants is an important task for government agencies and consulting firms in many parts of the world, and this is one of the principle employment opportunities for botany graduates!
2. It has been suggested, on the basis of molecular phylogenetics and statistical analysis of possible phylogenetic trees, that fused carpels (syncarpy) are the original condition of monocot flowers and that apocarpy evolved several times later (Sokoloff et al. 2013). This hypothesis, however, is challenged by the application of evolutionary, principles such as adaptation along the lines of least resistance and conservation of organization (Stebbins 1974). While adaptive pressures for fusion of carpels are several and strong, hypothesized pressures for splitting of carpels that are already fused together, particularly at early stages of development, are few and weak. Fusion of carpels brings stigmas together, making fertilization of all the ovules in the fruit more likely from a single pollinator visit, and also allows for better protected ovules and more diverse fruit types. Splitting into separate carpels might promote pollination of different carpels by different pollinators and more genetic diversity. The same result could be achieved however by producing more flowers with fewer ovules in each pistil.

GLOSSARY

acaulescent: literally "without a stem;" referring to a plant in which the primary shoot has very short internodes, or virtually none

achene: a single-seeded, hard, dry fruit in which the seed is attached at only one point to the fruit wall; in the narrow sense, derived from a single carpel, but also used for the dry fruit of the Asteraceae (technically a cypsela), which has a single seed in an ovary derived from two fused carpels; compare with nut and grain

actinomorphic: refers to flowers that are radially symmetrical; compare with zygomorphic

adaptation: an altered structural or physiological trait that increases the odds of survival, or the evolutionary process of such a trait coming into existence and becoming established in a population

adventitious bud: a bud that arises from a root, leaf, or older part of a stem, not formed during primary growth in the axil (at the base) of a leaf

adventitious root: a root that arises from a stem or leaf, rather than from a preexisting root

adventitious root system: refers to the collective root mass of a plant in which each root arises adventitiously from the stem, which is typically underground; also called a fibrous root system; as in the roots that form on clonal rhizomes, corms, and bulbs; compare with axial root system

aerenchyma: parenchyma tissue perforated with large intercellular spaces that are often organized into long canals; functions for buoyancy in floating aquatic leaves or whole plants, or for conducting air downwards to rhizomes and roots

aerial shoot: an upright photosynthetic and/or reproductive shoot in a plant that also has a permanent underground stem system

aerobic respiration: the process for releasing energy from stored foods using oxygen for complete breakdown

algae: eukaryotic organisms, other than terrestrial plants, that possess permanent, heritable chloroplasts

alternation of generations: a life cycle that consists of alternating haploid and diploid phases

anatropous: referring to an ovule that is bent over, as in most angiosperms; compare with orthotropous

annual herb: a nonwoody plant that completes its life within a single growing season

antenna complex: the array of pigments that gathers light in a photosystem

antheridium: a simple chamber on a gametophyte plant in which sperm cells develop

apical meristem: the small region of embryonic cells at the tip of a stem or root where new tissues and organs are generated

apocarpous: with carpels remaining separate from one another; not fused into a single compound pistil

archegonium: a chamber on a gametophyte in which a single egg develops

aril: a fruit-like layer on a seed

ascidiate: urn-shaped, with only a small opening at the top rather than an elongate suture, as in most carpels of the ANITA grade; compare with plicate

ATP: adenosine triphosphate; the primary energy courier within living cells

ATP synthase: molecular complex on internal cellular membranes through which protons flow, driving the synthesis of ATP

axial root system: the collective root mass of a plant that develops entirely through branching from the primary root of the embryo or a replacement root system of the same form; characteristic of most upright, nonclonal plants; compare with adventitious root system

axis: the central line around which an object can actually or theoretically rotate; in plants, the vertical center of an organ or the whole plant

bacteriochlorophyll: the form of chlorophyll found in photosynthetic bacteria other than the cyanobacteria

basal intercalary meristem: a narrow band of embryonic tissue at the base of an internode or leaf, by which these organs grow in length

berry: a fleshy fruit with one or more embedded seeds, without a hard pit

biennial herb: an herbaceous plant that accumulates food reserves during the first year of growth, then blooms and dies during the second season

bifacial leaf: a leaf in which the two sides of the leaf develop as distinct upper and lower surfaces; compare with unifacial leaf

bifacial vascular cambium: a vascular cambium that alternately produces xylem tissues toward the interior and phloem tissues toward the exterior of the plant

bilateral symmetry: having left-right, mirror-image symmetry, as in an orchid flower; zygomorphic; compare with radial symmetry

blade: the broad, flat portion of a leaf

bract: a flat, leaf-derived structure adapted to serve as a protective shield around tender tissues and buds, or when brightly colored to participate in the attraction of pollinators

brood-space pollination: transfer of pollen from one flower to another by adult insects that have developed from eggs laid within or among the immature flowers; as in the fig

bulb: a compact underground storage structure consisting of overlapping leaves or leaf sheaths modified for food storage; as in an onion

C_4 photosynthesis: a process that avoids photorespiration by physically separating the cells in which the Calvin cycle is occurring from the cells in which oxygen-generating light reactions are occurring

Calvin cycle: the metabolic pathway in which carbon dioxide is attached to preformed organic molecules, energized, and converted to carbohydrate

Cambrian period: 541–485 million years before the present; period in which major groups of animals evolved in the sea and atmospheric oxygen reached modern levels

capillary action: the upward movement of water through narrow spaces due to the cohesion of water molecules to one another and their adhesion to the walls of the spaces

Carboniferous period: 359–299 million years before the present; period of vast swamp forests dominated by seedless trees and early gymnosperms

carotenoids: red, orange, or yellow pigments; those in the antenna complexes of chloroplasts absorb light energy and transmit it to chlorophyll a

carpel: the enclosed chamber within which ovules develop; often differentiated into a distinct ovary, style, and stigma

carrion flowers: flowers adapted to resemble a dead animal in both smell and appearance; attract corpse-feeding insects that then pollinate the flowers

catkin: an elongate, flexible, pollen-bearing strobilus in wind-pollinated plants

cavitation: a rupture in the water column in a woody plant, introducing an air bubble that blocks further water transport; occurs when tension overcomes the cohesive properties of water due to excess evaporation at the top and/or inadequate water supply in the soil

chemosynthesis: a process of creating carbohydrate using energy from mineral sources; alternately referred to as chemoautotrophy

chlorophyll a: the primary pigment involved in the capture of light energy and formation of excited electrons in cyanobacteria and chloroplasts

chlorophyll b: a supplementary form of chlorophyll found in green algae and land plants

chloroplast: an organelle within eukaryotic cells that conducts photosynthesis; evolutionarily derived from a cyanobacterium

chromosomes: elongate structures consisting of a sequence of genes transmitted from generation to generation; in prokaryotes, the genome is arranged in a single circular chromosome; in eukaryotes, it is divided into several to many linear chromosomes, which typically condense during nuclear division into doubled, X-shaped structures

clonal: an organism that multiplies asexually; most commonly in plants via underground stem systems that extend outward and produce adventitious roots

coenocyte: an organ or entire organism functioning essentially as a single, multinucleate cell; as in the green algae *Acetabularia* and *Caulerpa*

coevolution: where the evolutionary histories of two kinds of organisms are closely intertwined, with changes in one resulting in changes in the other; as in the relationship between flowers and pollinating animals

cohesion: the attraction of water molecules to one another due to electrical charges

compound flower: a compact inflorescence adapted to resemble a single flower; as in the flower heads of the Asteraceae

compound leaf: a leaf that is subdivided into smaller leaflike units

conjugation: sexual union of cells from one individual to another via a tube that develops between them; as in the green alga *Spirogyra*

contractile roots: adventitious roots emerging from the base of bulbs and corms that contract to pull the plants deeper into the soil

convergent evolution: the evolution of similar structures and growth forms in unrelated organisms; a common example is the formation of a leafless, enlarged, succulent stem in families such as the Cactaceae, Euphorbiaceae, and Asclepiadaceae

corm: a short, upright, underground stem, usually roundish and wider than high, with one or more buds on top; new corms form at the base of developing buds each season as the older corms disintegrate; common in the Iridaceae

cotyledons: the one or two (sometimes more) leaves in a seed plant embryo; during germination, they may emerge above ground and photosynthesize or remain below ground to absorb nutrients stored in the seed

crassulacean acid metabolism: the process of storing carbon dioxide during the night and releasing during the day for use in the Calvin cycle; conserves moisture as stomata open for gas exchange only during the night

Cretaceous period: 145–66 million years before the present; final of the three periods of the Mesozoic Era; dominated by gymnosperms and early angiosperms; dinosaurs became extinct at the end of this period

cupule: a protective envelope around one or more ovules; common among seed ferns; evolved from leaf segments

cuticle: a layer of cutin: a hard, chemically inert polymer, deposited over the epidermis of exposed plant organs

cyclic electron flow: see cyclic photophosphorylation

cyclic photophosphorylation: process in which excited electrons drive the formation of ATP, then return to chlorophyll molecules

cytochrome complex: a pigment/protein complex where the inward flow of electrons draw protons across the thylakoid membrane into chambers for temporary storage

cytoskeleton: the complex system of rods and tubules within eukaryotic cells, which enables movement of organelles, chromosomes, and vesicles within the cell; also enables shape-changing and endocytosis in some cells

dark reactions: the light-independent reactions of photosynthesis; the Calvin cycle

deciduous leaves: leaves adapted to separate from the plant under specific environmental condition, such as a cold snap in autumn; chlorophyll is often first dismantled, leaving behind yellow to red accessory pigments

desiccation: the loss of water from cells

determinate: a growth pattern that has a predetermined endpoint; as in the maturation of animals or dicotyledonous leaves, the development of a flower, or of a shoot that ends in a flower or other reproductive structure

Devonian period: 419–359 million years before the present; major period of plant evolution in which early vascular plants evolved a diversity of growth forms including trees and the first seeds

dichotomous: a branching pattern consisting of repeated forking; as in the stems of early polysporangiophytes or the leaf venation of *Ginkgo* and many ferns

diploid: refers to a cell containing two complete sets of chromosomes or to a multicellular individual in which all the cells have two complete sets of chromosomes, one from each of its parents; compare with haploid

disk flower: in the compound flower heads of the Asteraceae, the radially symmetrical flowers with short petal tips usually occupying the center of the head; in some members of the family, all the flowers in a head are disk flowers; compare with ray flowers

dispersal: the movement of spores, pollen, seeds, or juvenile individuals within or between populations; serves to keep populations genetically mixed

double fertilization: process in angiosperms where the first sperm cell unites with the egg and the second unites with two nuclei of the central cell to initiate triploid endosperm

drupe: a fleshy fruit with a single large seed surrounded by a hard pit

electron carrier: molecules in the electron transport chain capable of briefly holding an excited electron, i.e., that can be readily reduced and oxidized

electron transport chain: the series of membrane-bound molecules that lower the energy of excited electrons in steps, one of which drives protons into the thylakoid; in noncyclic electron flow, it may also lead to the formation of NADPH

endosperm: the triploid food storage tissue in the angiosperm seed

endosymbiosis: the process by which bacteria or chloroplasts are captured and transformed into cellular organelles

epidermis: the outer covering of cells in plant organs; generally waterproofed with a cuticle and layers of wax; includes stomata and various hairs, scales, and other trichomes

epiphyte: a plant that lives on another plant; as in orchids and bromeliads that live on tree branches

eukaryotic cells: cells with nuclei, cytoskeleton, and membrane-bound organelles; characteristic of protists, plants, fungi and animals

eustele: the organization of the primary vascular bundles of a stem into a simple ring; characteristic of seed plants other than the monocots

flagellum: elongate filaments that provide locomotion to many unicellular organisms; consisting of simple rods of protein that spin like propellers in prokaryotes, and complex organelles with interior microtubules that wave back and forth in eukaryotes

floral tube: the funnel-like lower portion of certain flowers, formed from the fused bases of the petals and surrounding the reproductive organs; accumulates nectar at the bottom

flower: the bisexual reproductive strobilus of angiosperms; consisting of tepals, stamens, and carpels

follicle: dry fruit developing from an individual carpel that opens along the suture (or sometimes elsewhere) at maturity to release the seeds; conceptually the simplest and most ancient kind of fruit

fruit: the structure derived from the ovary portion of a pistil or individual carpel that houses the mature seeds; may be fleshy and edible or dry and hard or papery

fruit, false: fruit-like structures associated with mature seeds but which are not derived from the pistil or carpel; usually from the upper end of the flower stalk (receptacle); examples include apples and strawberries

fucoxanthin: a unique accessory pigment in the chloroplasts of brown algae, diatoms, and some other algae

gametes: haploid cells programmed to fuse together to create a zygote; commonly differentiated as sperm and egg

gametophyte: the haploid, gamete-bearing generation of a plant life cycle

generative cell: the cell within a pollen grain that divides to form two sperm cells

geophytes: perennial herbs that die off above ground and survive dormant periods as specialized underground structures; as in rhizomes, bulbs, corms, or tubers

germination: the process by which a spore or seed begins growth into a new plant; also applies to the beginning of pollen tube growth

gibberellin: the plant growth regulator (hormone) that is active in cell elongation

grain: a single-seeded dry fruit, similar to a nut but smaller, with the dry fruit wall tightly adhering to the seed throughout; technically, a caryopsis; characteristic of the grass family (Poaceae); compare with nut and achene

haploid: refers to cells or nuclei containing a single complete set of chromosomes

herb: a plant in which at least the above-ground portions are nonwoody and ephemeral; includes annual, biennial, and perennial forms

herbaceous: refers to nonwoody plants or plant parts

heterosporous: refers to a plant that produces two kinds of spores: megaspores and microspores

heterotroph: an organism dependent on obtaining organic energy resources from other organisms; applies to both the carnivorous habit of animals and the absorptive mode of bacteria and fungi

horizontal gene transfer: the acquisition and retention of foreign DNA by prokaryotes

hydroids: water-conducting cells in some mosses, lacking the strengthening layers of cellulose and lignin found in vascular plants

indeterminate growth: a pattern of growth characteristic of plants and fungi that can continue indefinitely, generally with branching to result in an ever expanding individual or colony

indusium: a flap-like covering over a cluster of sporangia (sorus) on the leaf of a fern

inflorescence: an aggregation of flowers on a specialized shoot, within which the leaves are reduced to bracts

integuments: the one or two jackets of cells around the nucellus of an ovule; evolved from leaf segments that surrounded ancient megasporangia; develop into the seed coat at maturity

intercalary growth: cellular division and/or elongation occurring between mature regions of a plant or plant organ; as in the elongation of the sporangium stalk in bryophytes, the elongation of internodes in a horsetail or bamboo, and the basal growth of a monocot leaf

intercalary meristem: a layer of embryonic cells within an internode or at the base of a leaf responsible for upward growth of these structures

internode: the stretch of stem between nodes

isogametes: gametes not differentiated as distinct sperm and egg; typically, both are motile and identical in size and appearance

kleptoplasty: the process of salvaging and exploiting chloroplasts from photosynthetic food items; the first step of secondary endosymbiosis

labellum: in orchids, the lower petal, which is elaborately colored and/or shaped to facilitate entry of a pollinating animal

landing platform: the lower, horizontally oriented petal or portion of a tubular flower, often decorated with contrasting stripes, dots, or hairs, serving as a platform on which bees land and enter the flower

leaf sheath: the broad lower portion of a leaf that envelopes the shoot apex and younger leaves; the broad base typically encircles the stem, connecting via a ring of parallel vascular bundles; characteristic of most monocots and some other plants

legumes: members of the family Fabaceae, most of which have mutualistic relationships with nitrogen-fixing bacteria

lichen: a mutualistic relationship between a fungus and an alga or cyanobacterium

life cycle: the phases of sexual reproduction, dispersal, and vegetative development that repeat from generation to generation

light reactions: the reactions involved in the capture of light energy and its use in synthesizing ATP and NADPH

light-independent reactions: the reactions of the Calvin cycle that produce carbohydrate, which are not driven directly by light energy

lignin: the strengthening polymer in the walls of tracheids, vessels, and fibers

LUCA: the last universal common ancestor from which all known life has descended

mangrove: a woody tree or shrub adapted to live in salt or brackish water

megagametophyte: a gametophyte that produces only eggs, not sperm cells

megaphyll: a complex leaf with multiple or branching veins; derived from a forking branch system; the type of leaf in all vascular plants except lycophytes

megasporophyll: a leaf or modified leaf bearing megasporangia or ovules

megaspores: spores from which egg-producing gametophytes will develop; compare with microspores

meiosis: nuclear division in a diploid cell that results in haploid daughter cells with different mixes of chromosomes from the previous generation

meristem: localized region of embryonic cells capable of division and expansion; source of all new tissues and organs; see apical and intercalary meristems and vascular cambium

Mesozoic era: approximately 252–66 million years before the present; the age dominated by reptiles, especially dinosaurs, and gymnosperms, with angiosperms appearing toward the end; consists of three periods: Triassic, Jurassic, and Cretaceous

microgametophyte: a gametophyte that produces only sperm cells, not eggs

microphyll: the simple, elongate leaf of a lycophyte, supplied by a single vein running through its length; derived from an outgrowth of the stem or from a sterilized sporangium; compare with megaphyll

micropyle: the small opening at the tip of an ovule through which a pollen tube will deliver sperm cells

microspores: spores from which sperm-producing gametophytes will develop; include pollen grains; compare with megaspores

microsporophylls: generally flattened structures that bear microsporangia or pollen sacs

midrib: the prominent central vein running through the length of a leaf

mimetic seed: a seed colored to resemble a fruit or aril, but without any fleshy tissue

mitosis: nuclear division that provides exact copies of the chromosomes in the mother cell to each daughter cell

monophyletic: a taxonomic group that consists of an entire clade: the common ancestor of the clade and all of its descendants; only monophyletic groups can be recognized in modern phylogenetic classification; compare with paraphyletic and polyphyletic

motor proteins: proteins in the cytoskeleton that move along the filaments and microtubules, facilitating the transport of vesicles and other objects

mucigel: slimy substance produced by a root cap to lubricate its movement through the soil

mutualism: a mutually beneficial symbiotic relationship between two different kinds of organisms

mycorrhizae: Specialized fungi living symbiotically with plant roots, aiding them in the absorption of water and mineral nutrients

NADPH: An energy transport molecule that incorporates high-energy electrons; produced in the noncyclic light reactions and provides energy to the Calvin cycle

nectar: a solution of sugar and other substances produced by many flowers for attracting animal pollinators

nectar spur: an elongate extension of a flower petal within which nectar collects

nectary: a gland from which nectar is produced

net venation: see reticulate venation

nitrogen cycle: the indefinitely repeated flow of nitrogen from the atmosphere through the ecosystem and back again

nitrogen fixation: the process in which atmospheric nitrogen is converted into ammonia
nitrogenase: the enzyme that splits nitrogen molecules (N_2) in biological nitrogen fixation
nocturnal: referring to events that occur at night
node: a point along a stem where a leaf and/or bud are attached; nodes occur regularly along a stem alternating with internodes
noncyclic electron flow: photosynthetic light reactions in which electrons are ultimately incorporated into molecules of NADPH
nucellus: the specialized sporangium within an ovule, containing a single small megaspore surrounded by food reserves
nut: a large, single-seeded fruit in which at least part of the fruit wall remains as a dry, hard, tightly adhering cover; as in an acorn; a walnut is only loosely called a nut, as only part of the fruit wall adheres to the seed; not peanut or brazil nut, which are both multiseeded; compare with achene and grain
orthotropous: referring to an ovule that is straight, as in gymnosperms; compare with anatropous
osmosis: the diffusion of water across a cell membrane from an area of high water concentration (fewer solutes) to one of low water concentration (more solutes)
ovule: an embryonic seed containing a specialized sporangium (nucellus) enveloped by one or two integuments; contains a single spore that develops into an egg-bearing gametophyte and, after fertilization, an embryo surrounded by food storage tissue
oxidation: the removal of electrons from one atom or molecule by another (see also redox potential)
pachycaul: technical term for a rosette tree, literally meaning "thick stem"
parallel venation: the pattern in which veins enter independently into the base of a leaf and remain separate and parallel to one another through the length of the leaf
paraphyletic: a formal taxonomic group that is not a complete clade; parts of the clade have been removed and assigned to formal taxa at the same rank; the traditional subclass Dicotyledonae was a paraphyletic group since the separate but equal Monocotyledonae also descended from its common ancestor; compare with monophyletic and polyphyletic
peptidoglycan: the durable polymer from which bacterial cell walls are built; consists of linked protein and carbohydrate units
perennial herb: a plant with ephemeral herbaceous shoots that persists year to year in the form of an underground stem/root system
petals: the inner series of tepals specialized for floral display; adapted for attraction of animal pollinators with different combinations of color, fragrance, and/or nectar
petiole: the narrow stalk of a leaf
phloem: tissue consisting of elongate cells specialized for conducting dissolved carbohydrate and other substances throughout the plant
photoautotroph: an organism that creates carbohydrate using light energy; i.e., a photosynthetic organism
photorespiration: a wasteful process in which carbohydrate is oxidized by rubisco; avoided by plants adapted for the C_4 metabolic pathway
photosynthesis: the process by which carbon dioxide and water are converted into carbohydrate, incorporating energy captured from sunlight
photosystem: the series of membrane bound pigments and protein complexes that capture light energy and convert it into ATP and NADPH
phycobilins: distinctive photosynthetic pigments found in cyanobacteria and red algae
pigments: molecules that absorb particular wavelengths of light, reflecting others; absorbed energy is dissipated as heat or as less energetic wavelengths of light; in plants, serve to provide color, diffuse ultraviolet radiation, or drive photosynthetic processes
pistil: the central, ovule-bearing structure of a flower, usually referring to one that consists of two or more carpels fused together, sometimes to the individual carpels when these remain separate
pits: thin areas in the thick cell walls of tracheids and vessels, through which water can flow
plankton: microscopic protists and animals that float freely in bodies of water

plate meristem: the collective embryonic cells undergoing division and expansion dispersed throughout the blade of an expanding dicotyledonous leaf

plicate: folded, as in a carpel with a distinct suture opposite its backbone (midrib), and with ovules lined up along the two edges of the suture; compare with ascidiate

pollen grains: the microspores of seed plants, within which a two- or three-celled microgametophyte develops, ultimately producing a pollen tube and two sperm cells

pollen sacs: chambers in which meiosis occurs, resulting in haploid pollen grains; equivalent to microsporangia; in angiosperms, an anther consists of a group of four elongate, parallel pollen sacs

pollen tube: the tubular outgrowth of a pollen grain that penetrates the stigma and style of a carpel or compound pistil, delivering sperm to an ovule

pollination: the process in which pollen is transferred from an anther to a stigma, usually between flowers on different plants; transfer may be accomplished by wind, water, or animal

pollinia: specialized anthers of orchids and some other plants in which pollen remains stuck together in masses that are transported intact from one flower to another

polyphyletic: a taxonomic group that includes species that did not descend from the same common ancestor; compare with monophyletic and paraphyletic

polysporangiophytes: the clade including the vascular plants plus ancient diploid sporophytes that did not yet have vascular tissue; characterized by sporophytes able to grow and branch via apical meristems and produce multiple sporangia

prickle: a sharp-tipped outgrowth of the outer tissues of a stem, as in a rose or blackberry; compare with thorn and spine

primary root: the first root of a plant, developing from the lower end of the root-stem axis; may persist to form a taproot or branch into an axial root system

primary thickening meristem: the region of a shoot apex, just below the apical meristem, in which cells increase more in width than in length resulting in a thick stem without secondary growth; common in monocots, particularly well developed in palms

primary vascular tissue: bundles of xylem and phloem produced during the growth of the shoot or root apex; compare with secondary vascular tissue

proboscis: a feeding structure of insects consisting of highly modified mandibles joined together to form a hollow tube; characteristic of Lepidoptera (moths and butterflies)

protists: eukaryotic organisms with relatively simple body organization, including unicellular, filamentous, or three-dimensional forms as in seaweeds; a grade of many kinds of eukaryotic organisms excluding plants, animals and fungi

proton motive force: the force of flowing protons from an area of high concentration to an area of low concentration; harnessed by cells to drive the synthesis of ATP

proton pump: a protein complex through which protons are forced across a membrane into restricted chambers, usually by a flow of electrons

protozoa: animallike protists that feed by engulfing other organisms

pseudobulb: an enlarged, water-storing section of an orchid stem

pseudocopulation: when a male insect mistakenly attempts to copulate with a flower adapted to resemble a female of the same species; a common means of pollination in orchids

pseudostem: a stemlike structure consisting of concentric leaf sheaths, each of which is longer than the preceding; as in a banana plant

radial symmetry: an organization in which body parts radiate from a central axis; actinomorphic; can be applied to whole plants and animals or portions thereof; as in flowers or rosettes of leaves; compare with bilateral symmetry

ray flower: a flower, located around the edge of a compound flower head in the Asteraceae, modified to resemble a single petal

rays: vertical wedges or sheets of parenchyma tissue running outward through the secondary xylem and phloem

receptacle: the tip of the flower-bearing stem to which the flower parts are attached; may be enlarged, flattened, concave, or extended; the edible portion of a strawberry is the portion of the receptacle to which the seedlike carpels are attached

reduction (redox) potential: a measure of the strength with which a compound or atom can pull electrons away from another substance; the receiving compound is thus reduced while the donating compound is oxidized

reticulate venation: netlike pattern of repeated, ever-finer branching and reconnecting of leaf veins characteristic of dicotyledonous plants, some gymnosperms and ferns, and a few monocots

rhizoids: in nonvascular plants, multicellular filaments that serve for anchorage and some external water uptake; in vascular plant gametophytes, threadlike extensions of individual cells that serve for anchorage and water absorption

rhizome: a horizontal, branching, underground stem that produces a series of adventitious roots and upright shoots

root cap: the thimble-shaped covering over the tip of a root; consists of cells produced by the root apical meristem that slough off as the root penetrates abrasive soil elements

root crown: the upper part of the root system where it connects with the shoot system; may refer to perennial herbs in which the root crown survives below ground, producing herbaceous stems each year

root-stem axis: the permanent midline of a vertical plant, running through the main stem and root, around which branch stems and roots radiate in a roughly radial symmetry; sometimes called the root-shoot axis; derived from a linear, bipolar embryo; as in most woody trees and shrubs

rosette: a large number of leaves attached close together at the tip of the stem and spread into a circular or mounded arrangement; as in a sundew or African violet

rosette tree: a thick-stemmed, rarely branching tree with a rosette of exceptionally large leaves around the shoot apex; as in a palm, cycad, tree fern or papaya plant; also called a pachycaul

rubisco: the enzyme involved in attachment of inorganic carbon to an organic carbon compound in the Calvin cycle; in the presence of high concentrations of oxygen can also oxidize carbohydrate (see photorespiration)

runner: usually refers to an above-ground stolon, as in a strawberry plant

sclerophyll: a small, hard, evergreen leaf adapted for water conservation and/or desiccation tolerance

scytonemin: a pigment protective against ultraviolet radiation found in some cyanobacteria

seaweed: a large, multicellular, plantlike alga

secondary endosymbiosis: the process by which chloroplasts are captured from other eukaryotic organisms and converted into permanent organelles; the starting point of several important groups of algae such as the brown algae and diatoms

secondary growth: production of secondary vascular tissues and bark in the stems and/or roots of woody plant resulting in increased thickness; also occurring in the taproots of plants like the carrot, resulting in layers of stored food tissue rather than wood

secondary phloem: phloem tissues produced by the vascular cambium; part of the bark

secondary plant compounds: compounds produced in plant tissues that are not used directly by the plant itself; evolved as adaptations that discourage animals from feeding upon them

secondary xylem: xylem tissue produced by the vascular cambium; accumulates to form wood

seed: a complex dispersal structure containing an embryo and stored food; the mature ovule

seed coat: the outer, usually waterproof covering of a seed, derived from the integument(s) of the ovule

seed leaf: a modified, seed-bearing leaf; a megasporophyll

sepals: the outermost, usually green tepals of a flower; sometimes modified into showy structures

shoot: a young section of stem with its leaves and buds

shoot apex: the tip of a shoot where new growth is occurring or may occur

shrub: a woody plant with multiple stems connecting to the root crown

Silurian period: 443–419 million years before the present; period in which the first land plants, including bryophytes and simple vascular plants, became established

sorus: a cluster of sporangia on the underside of a fern leaf

spadix: a dense spike of tiny flowers; as in the Araceae and some other families

spathe: modified leaf surrounding the spadix in the Araceae and some other families

sperm: male gamete, flagellate except in red algae, angiosperms, and some gymnosperms

spermatia: the tiny, nonmotile, current-dispersed male gametes of red algae

spine: a leaf or portion of a leaf modified into a sharp-tipped defensive structure; compare with thorn and prickle

sporangium: a multicellular chamber within which spores are produced; most often through meiosis

spores: specialized dormant and/or dispersal cells produced by bacteria, algae, fungi, and plants

sporophyte: the diploid, spore-producing generation in the plant life cycle

sporopollenin: the highly inert strengthening polymer in the walls of pollen grains and other spores

stamen: the characteristic pollen-bearing organ of flowering plants, consisting of a slender or flattened stalk (the filament) and four pollen chambers grouped together into an anther

stigma: the sticky tip of a carpel that captures pollen grains and stimulates the growth of pollen tubes

stolon: a slender, horizontal stem, below ground or on the surface, with long internodes and nodes that can root and produce a new plant

stomata: pores in leaves or photosynthetic stem tissues, bound by a pair of guard cells that can change shape to open or close the pore

strobilus: an aggregation of reproductive structures; most often cone-like; a flower is a modified strobilus, but flowers themselves can be aggregated into a strobilus-like structure

stromatolite: an upright rock formation consisting of thin layers of sediment alternating with the remains of photosynthetic bacterial mats; grows upward as bacterial mats are repeatedly buried and reestablished on the top

succulent: thick and filled with water-storage tissue; as in the thick leaves or stems of xerophytic plants

suture: the seam where the two edges of a carpel are joined together via interlocking cells

taproot: a single dominant root, usually thickened with food storage tissue or specialized for deep penetration of the soil

taproot system: an axial root system with a single dominant root present at least in the juvenile plant

tendril: a touch-sensitive, modified stem or leaf in a vine; adapted for wrapping around a supporting structure

tepals: flat, more-or-less leaflike structures surrounding the reproductive organs of a flower; most often in two series: sepals and petals

thallus: a vegetative structure with an indefinite shape, not organized into distinct stems and leaves; characteristic of some algae, liverworts, hornworts, and gametophytes of vascular plants

thorn: a short stem modified into a sharp-tipped defensive structure; compare with spine and prickle

thylakoids: membrane-bound chambers within chloroplasts and cyanobacteria where the light reactions occur; photosystems are embedded in the thylakoid membrane and hydrogen ions accumulate within the chamber

tracheid: a narrow, hollow water-conducting cell with pits for cell-to-cell movement; exclusive conducting cell in seedless vascular plants and most gymnosperms, supplemental in most angiosperms

transpiration: the upward movement of water due to capillary action and the evaporation and diffusion of water away from the plant

tuber: a specialized underground stem swollen with food storage tissue; as in a potato

tuberous root: a root, other than a taproot, thickened with food storage tissue; as in a sweet potato

turgor pressure: pressure that builds up within a plant cell due to inward osmotic diffusion and the restraint of the cell wall; responsible for the rigidity of soft tissues, cell growth, transport in the phloem, and leaf movements

unifacial leaf: a leaf that is developmentally folded with the two halves fused together, so that both sides are actually the lower side of the leaf; compare with bifacial leaf

unisexual: referring to flowers with either functioning stamens or functioning carpels, not both

vascular cambium: cylindrical layer of meristematic cells in woody plants; typically producing secondary xylem to the inside and secondary phloem to the outside

vascular tissues: xylem, phloem, and supporting tissues, typically running together within bundles or in separated layers in secondary growth

vein: a vascular bundle, typically reinforced with fibers, within the blade of a leaf

vesicle: a membrane-bound compartment functioning for storage or movement of materials within eukaryotic cells

vessel: a hollow, water-conducting tube made up of a series of empty cells, their walls strengthened with dense cellulose fibrils and lignin, and with open pores at the ends; found primarily in angiosperms

vine: a climbing plant, characterized by elongate internodes that either wind around supporting structures, or bear touch-sensitive tendrils for that purpose

whorl: refers to organs such as leaves in a circular arrangement around a stem; the organs of most flowers are in whorls as well

wood: the accumulated rings of secondary xylem in a stem or root

xerophyte: a plant adapted to dry environments; possessing succulent stems or leaves, or small, hard leaves (sclerophylls)

xylem: tissue consisting of hollow, water-conducting tracheids and/or vessels, plus supporting cells

zoospore: a flagellate cell produced by algae specialized for long-distance dispersal

zygomorphic: bilaterally symmetrical; usually referring to the overall structure of a flowers; contrasts with actinomorphic

zygote: the diploid cell resulting from the fusion of two gametes

BIBLIOGRAPHY

Algeo, T. J., S. E. Scheckler, and J. B. Maynard. 2000. "Effects of the Middle to Late Devonian spread of vascular land plants on weathering regimes, marine biota, and global climate." pp. 213–236. In: P. G. Gensel and D. Edwards (eds.). 2001 Plants Invade the Land: Evolutionary and Environmental Approaches. Columbia Univ. Press: New York.
Archangelsky, S., and R. Cuneo. 1990. *Polyspermophyllum*, a new Permian gymnosperm from Argentina, with considerations about the dicranophyllales. Review of Palaeobotany and Palynology 63: 117–135.
Asao, M., and M. T. Madigan. 2010. Taxonomy, phylogeny, and ecology of the heliobacteria. Photosynthetic Research, Springer online, 10.1007/s11120-009-9516-1.
Attenborough, D. 1995. The Private Life of Plants. Princeton University Press: Princeton, NJ. (see also the television series of this name, available as videorecordings)
Axelrod, D. I. 1952. A theory of angiosperm evolution. Evolution 6(1): 29–60.
Azuma, H., L. B. Thien, and S. Kawano. 1999. Floral scents, leaf volatiles and thermogenic flowers in Magnoliaceae. Plant Species Biology 14: 121–127.
Barth, F. G. 1991. Insects and Flowers: The Biology of a Partnership. Princeton University Press: Princeton, NJ.
Barthlott, W., S. Porembski, E. Fischer, and B. Gemmel. 1998. First protozoa-trapping plant found. Nature 392(6675): 447. doi:10.1038/33037.
Bethoux, O. 2009. The earliest beetle identified. J. Paleont. 83(6): 931–937.
Boyd, E. S., and J. W. Peters. 2013. New insights into the evolutionary history of biological nitrogen fixation. Frontiers in Microbiology. doi:10.3389/fmicb.2013.00201.
Butterfield, N. J. 2000. *Bangiomorpha pubescens* n. gen, n. sp: Implications for the evolution of sex, multicellularity, and the Mesoproterozoic/Neoproterzoic radiation of eukaryotes. Paleobiology 26(3): 386–404.
Chaloner, W. G., A. J. Hill, and W. S. Lacey. 1977. First Devonian platyspermic seed and its implications in gymnosperm evolution. *Nature* 265: 233–235.
Chaw, S.-M., C. L. Parkinson, Y.-C. Cheng, T. M. Vincent, and J. D. Palmer. 2000. Seed plant phylogeny inferred from all three plant genomes: Monophyly of extant gymnosperms and origin of Gnetales from conifers. PNAS 97(8): 4086–4091.
Chen, M., R. G. Hiller, C. J. Howe, and A. W. D. Larkum. 2005. Unique origin and lateral transfer of prokaryotic chlorophyll-*b* and chlorophyll-*d* light-harvesting systems. Molecular Biology and Evolution 22(1): 21–28.
Chen, P.-C., and C-L. Lai. 1996. Physiological adaptation during cell dehydration and rewetting of a newly-isolated *Chlorella* species. Physiologia Plantarum 96 (3): 453–457.
Cox, P. A., and R. J. Hickey. 1984. Convergent megaspore evolution and Isoetes. American Naturalist 24: 437–441.

Crow, J. F., and M. Kimura. 1965. Evolution in sexual and asexual populations. American Naturalist 99: 439–450.

Darwin, C. 1862. On the Various Contrivances by which British and Foreign Orchids are Fertilised by Insects. John Murray: London.

Davis, J. I., D. W. Stevenson, G. Peterson, O. Seberg, L. M. Campbell, J. V. Freudenstein, D. H. Goldman, C. R. Hardy, F. A. Michelangeli, M. P. Simmons, C. D. Specht, F. Vergara-Silva, and M. A. Gandolfo. 2004. A phylogeny of the monocots, as inferred from rbcL and atpA sequence variation, and a comparison of methods for calculating jackknife and bootsword values. Systematic Botany 29(3): 467–510.

Delwiche, C. F., L. E. Graham, and N. Thomson. 1989. Lignin-like compounds and sporopollen in *Coleochaete*, an algal model for land plant ancestry. Science 245(4916): 399–401.

Derenne, S., R. François, A. Skrzypczak-Bonduelle, D. Gourier, L. Binet & J.-N. Rouzaud. 2008. Molecular evidence for life in the 3.5 billion year old Warrawoona chert. Earth and Planetary Science Letters 272(1–2): 476–480.

Des Marais, D. J. 2000. When did photosynthesis emerge on earth? Science 289(5485): 1703–1705.

Dieringer, G., L. Cabrera R., M. Lara, L. Loya, and P. Reyes-Castillo. 1999. Beetle pollination and floral thermogenicity in *Magnolia tamaulipana* (Magnoliaceae). Int. J. Plant Sci. 160(1): 64–71.

Doyle, J. A. 2006. Seed ferns and the origin of angiosperms. J. Torrey Bot. Soc. 133: 169–209.

Doyle, J. A. 2013. Phylogenetic analyses and morphological innovations in land plants. Annual Plant Reviews 45: 1–50.

Drew, B. T, B. R. Ruhfel, S. A. Smith, M. J. Moore, B. G. Briggs, M. A. Gitzendanner, P. S. Soltis, and D. E. Soltis. 2014. Another look at the root of the angiosperms reveals a familiar tale. Systematic Biology 63: 368–382.

Endress, P., and J. Doyle. 2009. Reconstructing the ancestral angiosperm flower and its initial specializations. Am. J. Bot. 96(1): 22–66.

Endress, P. K., and A. Igersheim. 2000. Gynoecium structure and evolution in basal angiosperms. International Journal of Plant Sciences 161(Suppl.): 211–223.

Essig, F. B. 1971. Observations on pollination in *Bactris*. Principes 15: 20–24.

Essig, F. B. 1973. Pollination in some New Guinea palms. Principes 17: 75–83.

Faegri, K., and L. van der Pijl. 1979. The Principles of Pollination Ecology, 3rd. ed. Pergamon Press: Oxford.

Feild, T. S., N. C. Arens, J. A. Doyle, T. E. Dawson, and M. J. Donoghue. 2004. Dark and disturbed: a new image of early angiosperm ecology. Paleobiology 30: 82–107.

Field, T. S., T. Brodribb, and N. M. Holbrook. 2002. Hardly a relict: freezing and the evolution of vesselless wood in Winteraceae. Evolution (3): 464–478.

Foster, A. S., and E. M. Gifford. 1974. Comparative Morphology of Vascular Plants, 2nd ed. W. H. Freeman: San Francisco.

Friis, E. M., J. A. Doyle, P. K. Endress, and Q. Leng. 2003. *Archaefructus*: Angiosperm precursor or specialized early angiosperm? Trends in Plant Sciences 8: 369–373.

Frohlich, M., and M. Chase. 2007. After a dozen years of progress the origin of angiosperms is still a great mystery. Nature 450: 1184–1189.

Frohlich, M. W., and D. S. Parker. 2000. The mostly male theory of flower evolutionary origins: from genes to fossils. Systematic Botany 25: 155–170.

Gast, R. J., D. M. Moran, M. R. Dennett, and D. A. Caron. 2007. Kleptoplasty in an Antarctic dinoflagellate: Caught in evolutionary transition? Environmental Microbiology 9(1): 39–45.

Gerson, U. 1982. "Bryophytes and Invertebrates," Chap. 9. In: A. J. E. Smith (ed.). Bryophyte Ecology. Chapman & Hall: London and New York.

Glime, J. M. 2007. Bryophyte Ecology, Vol. 1. Physiological Ecology. Ebook sponsored by Michigan Technological University and the International Association of Bryologists. http://www.bryoecol.mtu.edu/

Graham, L. E., and L. Wilcox. 2000. Algae. Prentice Hall: Upper Saddle River, NJ.

Graham, S. W., J. M. Zgurski, M. A. Mcpherson, D. M. Cherniawsky, J. M. Saarela, E. S. C. Horne, S. Y. Smith, W. A. Wong, H. E. O'Brien, V. L. Biron, J. C. Pires, R. G. Olmstead, M. W. Chase, and H. S. Rai. 2006. Robust inference of monocot deep phylogeny using an expanded multigene plastid data set. Aliso 22(1): 3–20.

Grewe, F., W. Guo, E. A. Gubbels, A. K. Hansen, and J. P. Mower. 2013. Complete plastid genomes from *Ophioglossum californicum, Psilotum nudum,* and *Equisetum hyemale* reveal an ancestral land plant genome structure and resolve the position of Equisetales among monilophytes. BMC Evol Biol. 13(Jan. 11): 8. doi: 10.1186/1471-2148-13-8.

Halle, F. 2002. In Praise of Plants. Timber Press: Portland, OR.

Hemp, J., and L. A. Pace. 2010. Evolution of aerobic respiration. Presented at the Astrobiology Science Conference. League City, TX.

Hernández, Á. 2009. Birds and guelder rose *Viburnum opulus*: Selective consumption and dispersal via regurgitation of small-sized fruits and seeds. Plant Ecology 203(1): 111–122.

Hilton, J., and R. M. Bateman. 2006. Pteridosperms are the backbone of seed-plant phylogeny. J. Torrey Botanical Club 133(1): 119–168.

Ingrouille, M. J., and B. Eddy. 2006. Plants: Diversity and Evolution. Cambridge University Press: Cambridge, UK.

Jones, V. A., and L. Dolan, 2012. The evolution of root hairs and rhizoids. Annals of Botany: Ann Bot (2012); doi: 10.1093/aob/mcs136.

Kaiser, Roman A. 1993. The Scent of Orchids—Olfactory and Chemical Investigations. Elsevier: Amsterdam.

Kellogg, E. A. 2001. Evolutionary history of the grasses. Plant Physiology 125(3): 1198–1205.

Kenrick, P., and P. Crane. 1997. The Origin and Early Diversification of Land Plants: A cladistic study. Smithsonian Institution Press: Washington, DC.

Kim, E., and L. E. Graham. 2008. EEF2 analysis challenges the monophyly of Archaeplastida and Chromalveolata. PLoS ONE 3(7): e2621. doi:10.1371/journal.pone.0002621.

Knoll, A. 2014. Paleobiological perspectives on early eukaryotic evolution. Cold Spring Harb. Perspect. Biol. 2014(6):a016121.

La Farge, C., K. H. Williams, and J. H. England. 2013. Regeneration of Little Ice Age bryophytes emerging from a polar glacier with implications of totipotency in extreme environments. Proceedings of the National Academy of Sciences. Published online May 27, 2013. doi:10.1073/pnas.1304199110.

Lee, E. K., A. Cibrian-Jaramillo, S. Kolokotronis, M. S. Katari, A. Stamatakis, M. Ott, J. C. Chiu, D. P. Little, D. W. Stevenson, W. R. McCombie, R. A. Martienssen, G. Coruzzi, and R. DeSalle. 2011. A functional phylogenetic view of the seed plants. PLOS Genetics. doi:10.1371/journal.pgen.1002411.

Ligrone, R., J. G. Duckett, and K. S. Renzaglia. 2012. The origin of the sporophyte shoot in land plants: a bryological perspective. Annals of Botany, published online. doi:10.1093/aob/mcs176.

Margulis, L. 1998. Symbiotic Planet. Basic Books: New York.

Markham, K., S. Bloom, R. Nicholsonb, R. Rivera, M. Shemluck, P. Kevan, and C. Michener. 2004. Black flower coloration in wild *Lisianthius nigrescens*: Its chemistry and ecological consequences. Z. Naturforsch. 59: 625–630.

McAuslane, H. J. 2012. University of Florida IFAS extension. http://edis.ifas.ufl.edu/in135.

Meeuse, B. J. D. 1961. The Story of Pollination. Ronald Press: New York.

Menand, B., K. Yi, S. Jouannic, L. Hoffmann, E. Ryan, P. Linstead, D. G. Schaefer, and L. Dolan. 2007. An ancient mechanism controls the development of cells with a rooting function in land plants. Science 316(5830): 1477–1480.

Mitton, J. B., and M. C. Grant. 1996. Genetic Variation and the Natural History of Quaking Aspen, BioScience 46 (1): 25–31.

Niklas, K. 1997. The Evolutionary Biology of Plants. University of Chicago Press: Chicago, IL.

Nisbet, E. G., N. Grassineau, C. J. Howe, and R. E. R. Nisbet. 2007. The age of rubisco: the origin of oxygenic photosynthesis. Geobiology 5: 311–335.

Norstog, K. 1987. Cycads and the origin of insect pollination. American Scientist 75: 270–278.

Nozaki, H., S. Maruyama, M. Matsuzaki, T. Nakada, S. Kato, and K. Misawa. 2009. Phylogenetic positions of Glaucophyta, green plants (Archaeplastida), and Haptophyta (Chromalveolata) as deduced from slowly evolving nuclear genes. Molecular Phylogenentics and Evolution 53:872–880.

Pierce, S. K., J. J. Hanten, S. E. Massey, and N. E. Curtis. 2003. Horizontal transfer of functional nuclear genes between multicelllular organisms. Biol. Bull. 204: 237–240.

Pisciotta, J. M., Y. Zou, and I. V. Baskakov. 2010. Light-dependent electrogenic activity of cyanobacteria. PLoS One 5(5): e10821.

Prestianni, C., Y. Caraglio, Y. Guedon, B. Meyer-Berthaud, and P. Gerrienne. 2007. The seed plant *Moresnetia:* Morphology and anatomy of the stem and fertile frond. Abstract, Botany & Plant Biology Joint Congress 2007.

Proctor, M., P. Yeo, and A. Lack. 1996. The Natural History of Pollination. Timber Press: Portland, OR.

Rasmussen, B., I. R. Fletcher, J. J. Brocks, and M. R. Kilburn. 2008. Reassessing the first appearance of eukaryotes and cyanobacteria. Nature 455: 1101–1104.

Raven, J. A., and D. Edwards. 2001. Roots: evolutionary origins and biogeochemical significance. Journal of Experimental Botany 52(90001): 381–401.

Raymond, J., J. L. Siefert, C. R. Staples, and R. E. Blankenship. 2004. The natural history of nitrogen fixation. Mol. Biol. Evol. 21(3): 541–554.

Raymond, J., O. Zhaxybayeva, J. Gogarten, S. Gerdes, and R. Blankenship. 2002. Whole-genome analysis of photosynthetic prokaryotes. Science 298(5598): 1616–1620.

Renner, S. 2009. "Gymnosperms," Chap. 15. In: S. B. Hedges and S. Kumar (eds.). The Timetree of Life. http://www.timetree.org/pdf/Renner2009Chap15.pdf

Renzaglia, K. S., R. J. T. Duff, D. L. Nickrent, and D. J. Garbary. 2000. Vegetative and reproductive innovations of early land plants: implications for a unified phylogeny. Philos. Trans. R. Soc. Lond. B Biol. Sci. 355(1398): 769–793.

Retallack, G., and D. L. Dilcher. 1981. Arguments for a Glossopterid ancestry of angiosperms. *Paleobiology* 7(1): 54–67.

Reyes-Prieto, A., A. P. M. Weber, and D. Bhattacharya. 2007. The origin and establishment of the plastid in algae and plants. Annu. Rev. Genet. 41: 147–168.

Robertson, R. E., and S. C. Tucker. 1979. Floral Ontogeny of *Illicium floridanum*, with emphasis on stamen and carpel development. Amer. J. Bot. 66(6): 605–617.

Rumpho, M. E., J. M. Worfula, J. Leeb, K. Kannana, M. S. Tylerc, D. Bhattacharyad, A. Moustafad, and J. R. Manharte. 2008. Horizontal gene transfer of the algal nuclear gene psbO to the photosynthetic sea slug *Elysia chlorotica*. PNAS 105(46): 17867–17871.

Schlesinger, W. H. 1991. Biogeochemistry: an analysis of global change. Academic Press. San Diego, CA.

Schopf, J. W. 2006. Fossil evidence of Archaean life. Phil. Trans. R. Soc. B 361: 869–885.

Sokoloff, D. D., M. V. Remizowa, and P. J. Rudall. 2013. Is syncarpy an ancestral condition in monocots and core eudicots? In: Early Events in Monocot Evolution, P. Wilkin & S. J. Mayo (eds.). Cambridge University Press: Cambridge, UK.

Stebbins, G. L. 1974. Flowering Plants: Evolution Above the Species Level. Belknap Press, Harvard University Press: Cambridge, MA.

Stevens, P. F. 2001 onwards. Angiosperm Phylogeny Group (http://www.mobot.org/MOBOT/Research/APweb/).

Stevenson, D.W., Norstog, K. J., and Fawcet, P. K. S. 1998. "Pollination biology of cycads," pp. 277–294. In: S. J. Owens and P. J. Rudall (eds.). Reproductive Biology. Royal Botanic Gardens: Kew.

Stoecker, D. K., A. E. Michaels, and L. H. Davis. 1987. Large proportion of marine planktonic ciliates found to contain functional chloroplasts. *Nature* 326: 790–792.

Strother, P. K., S. Al-Hajri, and A. Travere. 1996. New evidence for land plants from the lower middle Ordovician of Saudi Arabia. Geology 24: 55–58.

Summons, R. E., L. L. Jahnke, J. M. Hope, and G. A. Logan. 1999. 2-Methylhopanoids as biomarkers for cyanobacterial oxygenic photosynthesis. Nature 400: 554–557.

Sun, G., Q. Ji, D. L. Dilcher, S. Zheng, K. C. Nixon & X. Wang. 2002. Archaefructaceae, a new basal angiosperm family. Science 296(5569): 899–904.

Taylor, D. W., and G. Kirchner. 1996. "The origin and evolution of the angiosperm carpel," pp. 116–140. In: D. W. Taylor and L. J. Hickey (eds). Flowering Plant Origin, Early Evolution and Phylogeny. Chapman & Hall: New York.

Tomescu, A. M. F. 2008. Megaphylls, microphylls and the evolution of leaf development. Trends in Plant Science 14(1): 5–12.

Tomlinson, P. B., and B. A. Huggett. 2012. Cell longevity and sustained primary growth in palm stems. American Journal of Botany 99(12): 1891–1902.

Turlings, T. C. J., J. H. Tumlinson, and W. J. Lewis. 1990. Exploitation of herbivore-induced plant odors by host-seeking parasitic wasps. Science 250: 1251–1253.

van der Pijl, L., and C. H. Dodson. 1966. Orchid Flowers: Their Pollination and Evolution. The Fairchild Tropical Garden and the University of Miami Press: Coral Gables, FL.

Vasek, F. C. 1980. Creosote bush: Long-lived clones in the Mojave Desert. Amer. J. Bot. 67: 246–255.

Wang, X., and G. Han. 2012. The earliest ascidiate carpel and its implications for angiosperm evolution. Acta Geologica Sinica, English Edition 85(5): 998–1002.

Waterbury, J. B., S. W. Watson, R. R. L. Guillard, and L. E. Brand. 1979. Widespread occurrence of a unicellular, marine, planktonic cyanobacterium. Nature 277: 293–294).

Willis, K. J., and J. C. McElwain. 2002. The Evolution of Plants. Oxford University Press: Oxford, UK.

Wu, C-S., Y-N. Wang, S-M. Liu, and S-M. Chaw. 2007. Chloroplast genome (cpDNA) of *Cycas taitungensis* and 56 cp protein-coding genes of Gnetum parvifolium: Insights into cpDNA evolution and phylogeny of extant seed plants. Molecular Biology and Evolution 24: 1366–1379.

Xue, C., R. Huang, T. J. Maxwell, and Y. Fu. 2010. Genome changes after gene duplication: haploidy vs. diploidy. Genetics 186(1): 287–294.

Yuan, X., S. Xiao, and T. N. Taylor. 2005. Lichen-like symbiosis 600 million years ago. Science 308(5724): 1017–1020.

Zimmer, C. 2009. On the origin of sexual reproduction. Science 324(5932): 1254–1256.

INDEX

Note: page entries that contain illustrations are in italics.

Abrus precatorius, 159
Acacia, *198*, 199–200
Acaulescent, *171*, 172
Acetabularia, 40, 42
Achenes, 162, 163
Acorus, 224, 226, 227, 228
Adaptive modification (Stebbins principles), 134, 236
Adenia, 188
Adventitious buds, 174–175
Adventitious roots, 91, 101, 172, 173, *174*, 175, 200, *211*, 213
See also root systems, adventitious
Aerenchyma, 193, 218, 220, 223
Aerial shoots, 74, 75, 80, 81, 84
Aerobic respiration, 4, 15, 21
Agave, 201, 216, 218, 219
Aglaophyton, 66, 67, 74–75
Aldrovanda, 190–191
Algae, xiv–xv, 1, 20, 21, 22, 23, 29
  as endosymbionts 17, 18, *19*, 25, 27
  brown, 23, 25, *26*, 28
  green, 25, 27–32, *36–45*, 50–53, 62
  red, 24, 25, 27–28, 29, 30–32, 63
Alismataceae, 222, 223
Alismatales, 218, 224, 226, 228
Alliaceae, as source of medicinal compounds, 196
Alnus, 156
Aloe, 146, 215, 216, *219*
Alternation of generations, 38
  in algae, 37, 64
  in land plants, 61–63, 67, 68
Amaryllidaceae, 205
Amaryllis, *204*, 205, 212
Amborella, 131–132, *133*, 135, 171, 192, 224, 235
Amborellales, 133, 235

Amorphophallus titanum, 153
Angiosperms, xii, 110, 113, 116–135, 156, 159–160, 162, 167, 176, 182, 205,235, 236
Angraecum sesquipedale, *136*, 137–138, 141–142
Animals, photosynthetic, *16*, 17–18, 27, 29
ANITA grade, 131–132, *133*, 135, 167, 192, 235, 236
Annual herbs, 171–172, *173*
Antenna complex, 8, *10*, *11*
Antheridia, *61*
Anthocerophyta, 48
Anthoceros, 48, *49*, 69
Anthurium, 152, 212
Antithetic hypothesis, 67
Apiaceae, 151, 171
Apocarpous flowers, in monocots, 228, 229, 236
Apocynaceae, 140, 188
Aquatic plants, 236
  in bryophytes, 53
  in seedless vascular plants, 81, *83*, 96, 98
  in dicotyledonous plants, 118,119, 132, 157, 192, 193, *194–195*
  in monocots, *159*, 218–219, *220–223*, 229
Aquificales, 15
Aquilegia, 145
Araceae, 147, 151, *152*, 156, 207, 212, 218, 222, 228
Araucaria, 106
Archaea, 15, 20, 235
Archaefructus, 119, 122
Archaeopteris, 89, *91*, 103, 104
Archaeplastida, 27, 29
Archegonia, *61*, *62*, 97, 99
Aril
  in angiosperms, 135, 159
  in gymnosperms, 95

Arisaema, 156
Aristolochia, 156
Aristolochiaceae, 147, 177, 224
Asarum, 224
Asclepiadaceae, 147, 148
Asteraceae, 151, 153, 188, 193
ATP, 5, 7, 9, 10, 11, 13
ATP-synthase, 10
Axis, root-stem, 91, 92, 101, 172
Azolla, 98

Bacteriochlorophyll, 10–13
Bactris, 149–151
Bamboos, 83, 89, 160, 201, 213, 215, *217*
Banded iron formations, 3–4
Bangiomorpha pubescens, 27, 29
Banksia, 146
Basal intercalary meristem
   in *Welwitschia* leaves, *109*
   in monocot leaves, 199, 202–203, *204*, 205, 224
   in bamboo internodes, 215
Bats, as pollinators, 140, 146, *147*, 149
Bees, as pollinators, 139, 140–145
Beetles, as pollinators, 113, 147, 150, *151*
Begonia, 177
Bennettitales, 122, 127, *128*
Berry, 134, 135, 159, 160–162, 236
Betula, 156
Bidens, 193
Biennial herbs, *171*, 172
Bifacial leaf, 228
   vascular cambium, 89, *90*
Bilateral symmetry, 87
   in flowers, *144*
Birds and seed dispersal, 95, 135, 159, 161–162
   as pollinators, 117, 140, 145, *146*
Blue-green algae, 1
Borrichia, 189
Brassicaceae, 160
Brighamia, 141–142
Bromeliads, as carnivorous plants, 189
Bromeliads, as epiphytes, 201, 212–213, *214*
Brood-space pollination, *154*, 155
Broussonetia papyrifera, 174
Brown algae. See algae
Bryophytes, xv, 46, 48, *49*, 50–66
   size limitation, 69–70
Bulbs, 201, 208, *212*
Bumblebees, as pollinators, nectar thieves, 143
Butomus, 226, *228*
Butterflies, as pollinators, 137, *138*, 141

C_4 photosynthesis, 217
Cabomba, 193, *195*, 224
Cacti, 57, 58, *147*, 161, *187*, 188
Calamites, 89, *90*

Calvin cycle, 6, 7, 8, 9, 10, 13
Calycanthus, 156
Cambrian explosion, 1
Canellaceae, 171
Capillary action, 54–56, 91
Carboniferous period, 74, 89, 93, 96, 102, 103, 104, 110, 113
Carnegia gigantea, 146
Carnivorous plants, 188–189, *190–192*
Carotenoids, 25, 30
Carpel, 118, *120*, *121*, 122–123, *124*, 125, 127, 130–132, *133*, 134–135, 139, 151, 159, 163, 227, *228*, 229, 236
Carrion flies, as pollinators, 147, 156
Casuarina, 156
Catkins, *105*, 107, *157*
Caulerpa, 40, 42
Cavitation, 170
Caytonia, Caytoniales, 127–128, *129*, 130–131
Cephalotaxus, 135
Cephalotus, 189
Ceratophyllum, 193, 224
Chara, 43, 44
Charophytes, 44, 51, 65, 69
Chemosynthesis, 6, 8, 13, 235
Chlamydomonas, 23, 36, 38, 41
Chlorella, 51
Chlorococcum, 51
Chlorophyll, 8, 9, 10, 11
   chlorophyll a, 12, 25, 30
   chlorophyll b, 25, 30
Chloroplasts, xv, 8, 9, 16–19, 21–23, 25–28, 30–31, 40, 41, 53, 58, 62
Chromosomes, 33–35
Ciliates, photosynthetic, 18
Cissus, 188
Clematis, 143, 163, *164*
Clonal growth, 48, 73–75, 175, 180, 200
   in trees, 172
Club mosses, 56, 77–79, *80*
Coal, deposition in Carboniferous period, 74, 89
Coenocytic green algae, 42
Coleochaete, 44, 45, 51, 52, 62–63, 69
Color
   in pollination, 114, 117, 121, 123, 139, 142–149, 152, 156, 184
   in fruit dispersal, 135, 159–162
Commelinaceae, 139
Compound flowers, *152–153*
Compound leaves, 183, *184–185*, 208–209, 225
Conifers, 96, 99–100, 103–105, *106–107*, 110, 113, 127, 135, 174, 186
Conjugation, in algae, 39, *40*
Contractile roots, 212
Convolvulaceae, 175
Cooksonia, 75, 84

Index

Cordaites, Cordaitales, *103*, 104–105
Corm, 79, 177, 210, *211*, 212
Cornaceae, 153
Coryanthes, 145
Corylus, 157
Corystospermales, 127–128, 130
Cotyledons, *101*, *168*, 201–202, *203*, 225, 226–227
Cradle environments, 118
Crassulaceae, 188
Crassulacean acid metabolism, 188, 217
Crown group, of angiosperms, *119*, 120, 122, 135
Cupule, *129*, *130*, 131, 134
Cuticle, 56–58, 188
Cyanobacteria, xiv, xv, 1, 2, 3–5, 8, 12, 13–15, 18–21, 25, 28, 30, 50
Cycads, 96, 98–100, 104, 110, *111–112*, 113–114
Cycas, 112
Cyclanthaceae, 151, 228
Cyclic electron flow, 9, *10*, 13
Cyperaceae, 219
Cyperus papyrus, 222, 223
Cytochrome complex, 9–11
Cytoskeleton, 20–23, 31, 33

Dahlia, 177, *180*
Dark reactions, 5
Darlingtonia, 189
Darwin, Charles, xiii, xiv, 117, 118, 136–138, 191
Dawsonia, 69
Deception in pollination, 147–149, 156
Deciduous leaves, 156, 169, 185–186
Decodon, 193
Degeneria, 124
Desiccation tolerance, 48, 50–51, 53, 56
Devonian period, 73–75, 89, 102, 103, 113
Diatoms, 25–26, 27, 40
Dicentra, 40
Dichotomous branching, 75, 102
  in leaf venation, 104
Dicotyledonous plants, 167–196
Dinoflagellates, 25, 27, *28*
Dioscorea, 196
Disk flowers, *153*, 154
Dispersal, 31, 35, 37, 52, 62, 63, 65, 95, 98, 100, 113, 158, 160–165
Double fertilization, 126, 130
Dracaena draco, 215, *218*
Drosera, 190, *191*
Drupes, 134–135, *161*, 162, 236

Ecballium elaterium, 160
Echites umbellata, 196
Eichhornia, 222, 223, 224

Egg, 31, 32, *39*, *44*, 48, *61*, 96–97, *100*, 126
Electron transport chain, 8, 9, 12, 13
Elkinsia, 102
Elodea, 219, 222, 224
Elysia chlorotica, *16*, 17
Embryo, 22, 48, *61*, 68, 92, 95–97, 99, *100*, *101*, 201, 225
Embryophyta, 48
Endosperm, 126, 202
Endosymbiosis, *21*, 22, 25–30, 231
Ephedra, 108, *109*, 196
Epidermis, 56, *58*, 74
Epiphyte, 85, 212, *213–214*, 231
Equisetum, 82, *84*, 86
Erica, 146
Eryngium yuccifolium, 225, 226–227
Eucalyptus, 73, 174
Eudicots, 162, 167, 170–171
Euglenids, 25, *28*
Euglossine bees, as pollinators, 145
Eukaryotes, 1, 4, 15, 19–23, 29, 235
Euphorbia, *187*, 188
Euphorbiaceae, 118, 153, 187
Euphyll, euphyllophytes, 76, 79

Fabaceae, 15, 159, 180
Ferns, 66, 79–81, 82, 83, 85, 87, 88, 91, 96, 97, 98, 100–101, 102, 104, 129, 132, 172, 182, 184, 196, 201, 212
  See also seed fern
Ficus, 154, 173, *174*
Flagella, 22, 23, 30–32, 33, 38–39, 44, 61–62, 98–100, 110
Flies, as pollinators, 139–140, 142, 147, 150, *151*, 155, 156
Flower, *120*, *121*, 122–126, 132, 134, 137–159, 161, 184
Flower, compound, *152–153*, 154
Follicles, 123, *124*
Forests, 45–48, 73–74, 89, 93, 166–167, 186, 201, 212, 231
Fossilization, xiv, 5, 102, 117–118,
Fragrance, in pollination, 114, 121, 123, 139, 142–143, 145–148, 156
Fruits, 125, 133–135, 158, *160–164*, 165
Fuchsia, 145, *146*
Fucoxanthin, 25
Fumaria, 140

Gametes, 32, 35, *36–37*, 38, 39
Gametophyte, *37–38*, 52, 60–61, 64–66, 68, 70, 74, 96, 97, 99, *100*, 110, 125–*126*
Genlisea, 191
Geophytes, 210
Gibberellin. 176–178
Ginkgo, ginkgophytes, 96, 98, 99, 100, 104, *105*
Glaucocystophytes, 21

Glossopteridales, 127–128, 130
Gnetophytes, 107–110, 127
Gnetum, 107, *108*, 182
Grasses, 107, 110, 118, 156, *158*, 164, 169, 196, 198, 201, 205–206, 210, 216–217, 219, 226, 228, 229, 231–232
  See also sea grasses
Green algae. See algae
Green gliding bacteria, 13
Green sulfur bacteria, 13
Grevillea, 146
Gunnera, 177

Hawk moths, as pollinators, *136*, 138, 141–142
Heliamphora, 189
Helicodiceros muscivorus, 156
Heliconia, 145
Heliobacteria, 13
Hemerocallis, 25
Herbaceous growth forms, 168, 169, 171–172, 173, 175–177, 193, 196, 198, 200–201, 206, 208, 210, 214
Heterosporous life cycle, 96–98
Heterotrophs, 4–6, 13
Hippuris, 193
Honeaters, as pollinators, 145
Honeycreepers, as pollinators, 145
Horizontal gene transfer, xiv, 12–13, 15
Horneophyton, 74, 75
Hornworts, xv, 45, 47–48, 49, 51, *52*, 53, 60–61, 64, 69
Horsetails, 81, *84*, 86, 88–89, 96, 175, 215
Hummingbirds, as pollinators, 145, *146*
Hura crepitans, 160
Hydriastele, 150, *151*
Hydrilla, 219, 224
Hydrocotyle, 176, 193
Hydrodictyon, 40, 43
Hydroids, 56, 60

Impatiens, 160
Indeterminate growth, 44, 48, 77, 203
Indusium, 81, *82*
Inflorescence, 122, 142, 147, 149–156, 172
Intercalary growth, 64, 83, 109, 175, 192, 202, 204, 205, 222–224
Iridaceae, 205, 210, 228
Iris, 118, 144, 177, *200*
Isoetes, 79, *81*, 96, 98

Juncaceae, 219
Juniperus, 135

Kleptoplasty, 18–19, 22, 23, 25, 29, 30

Lagarostrobos franklinii, 174
Lamiaceae, 140, 144
Larrea tridentata, 175, 188
Leaves
  in bryophytes, 48, 53, 54, 56
  in seedless vascular plants, 77–78, 79, *80*, 82, 83–84, *85*, 86, 88
  in gymnosperms, 101, *102–112*, 127, *128*
  in dicotyledonous plants, 167–169, 180–181, *182–186*, 188–189, *190*, *192*, *194*, *195*
  in monocot, 199–203, *204–209*, *212*, *219–223*, *224–225*, *226–227*, 228
Legumes, 15, 231–232
Lemna, 222
Lentibulariaceae, 191
Lepidodendron, 89, *90*
Lepidoptera, as pollinators, 137, *138*, 140–141
Lichens, *18*, 19, 50
Life cycles, 15, 33, *36–38*, 48, 52, 66, 96, 158
Light reactions of photosynthesis, 5, 8, *10–11*, 13
Lignin, 59, 60, 69–70, 75
Lilium, 212
Lisianthus nigrescens, 143
Liverworts, xv, 45, 47, *48*, 49, 51, *52*, 53, 60–61, 63, 64, 65–68
Lodoicea maldivica, 95
Lonicera sempervirens, 145
LUCA, 2, 15
Ludwigia, 193
Lycophytes, 76–78, 96, 98

Magnolia, *121*, 122, 124, 139, 142, 149–151
Magnoliales, Magnolids, 133–135, 162, 167, 170–171, 177, 192, 224–225, 235, 236
Malus, 140, 196
Mammals, as pollinators, 146, *147*
Mangroves, 103, 155, 173, 192, 215, 229
Marchantia, 48–49
Marchantiophyta, 48
Margulis, Lynn, 21, 23, 28
Marsilea, 98
Megaphylls, 78–79, 82
Megaspore, 96, 97, 99, *100*
Meiosis, 32–33, 35, 38
Meristem
  in carpels, 134–135
  apical, 48, 66, 68, 77, 80, 83, 182
  intercalary, 64, 83, 109, 202, *204*, 205, 224
  plate, 182
  primary thickening, 209, 215
  See also vascular cambium
Microspores, microgametophyte, 97, 98
Microphylls, 78, 79, *80*, 81
Mitosis, 33, *34*, 38
Monocots, 83, 133–135, 156, 167–169, 172, 177, 183, 192, 196, 198–229, 235, 236
Moresnetia, 102

Mosses, 46, 47–48, 49, 51, 53, 56, 60, 61, 64, 65, 67
   size limitation, 69–70
Moths, as pollinators, 136–138, 140–142, 151, 155
Musa, 206, 208
Museum habitats, 118, 127
Mutualism, 18, 29
Mycorrhizae, 77
Myriophyllum, 193, 195

NADPH, 5, 7, 10–11
Nectar, in pollination, 118, 123, 139–141
Nectar spur, 137, 142, 143–147, 151, 189
Nelumbo, 193, 194
Nepenthes, 189, 190
Nerium oleander, 196
Net venation, 181, 182
   in monocots, 206–207
Nitrogen fixation, cycle, 2, 13, 14, 15, 231
Non-cyclic electron flow, 10, 11
Nonvascular plants, 48, 54, 56, 66–69
Nuphar, 140
Nuts, 95, 163, 164, 165
Nymphaea, 121, 194
Nymphaeales, 118, 132–133, 140, 192, 195
Nymphoides, 193, 194
Nypa fruticans, 155, 215, 229

Oleaceae, 151
Ophioglossales, 85
Ophrys, 148, 149
Orchids, 95–96, 136, 137–138, 141–142, 144–145, 148, 149, 153, 156, 159, 169, 196, 201, 205, 212, 213
Ordovician period, 51
Osmosis, 55, 59
Ovule, 96, 99, 100, 101, 102, 103–104, 105, 108, 112
   in angiosperms, 118, 120, 122–123, 124, 125, 126, 129, 130, 131, 134
Oxygen buildup in oceans and atmosphere, 3–4, 39
Ozone layer, 8, 39, 50

Pachycaul, 184
Pachypodium, 188
Paederia foetida, 180
Paeoniaceae, 139
Paleoherb, 177, 192
Pandanus, Pandanaceae, 201, 215, 216, 228
Papaveraceae, 139, 140, 143
Parallel venation, 169, 183, 199, 202, 204, 209, 220, 224–225
Pelargonium, 188
Peperomia, 151–152, 224
Perennial herb, 173, 175–177, 198, 200, 208, 210, 214, 224

Permian period, 96, 102, 104, 105, 110
Philodendron, 152, 205, 207, 212
Phloem, 59, 66, 89, 90, 169, 215
Photoheterotroph, 13
Photorespiration, 217
Photosynthesis, 1–13
   summary equation, 4
Photosystems, 8–13
Phycobilins, 28, 30
Phycoerythrin, 24, 28
Picea abies, 174
Pigments and evolution of photosynthesis, 8
Pinus longaeva, 73
Piperaceae, 151, 152, 224
Pistia, 222
Pistil, 120, 124, 125
Pitcher traps, 189, 190
Plankton, 1, 2, 18, 39–40
Platanaceae, 193
Plate meristem, 182
Pleopeltis polypodioides, 56, 57
Poaceae, 201, 219
Podocarpus, Podocarpaceae, 106, 135, 174
Poinsettia, 123, 153, 184
Pollen, 32, 62, 63, 66, 74, 93, 96–98, 99, 100, 113–114
Pollen sacs, 98, 101, 105, 114, 123, 124, 128, 129
Pollen tube, 99, 123, 125, 127, 131
Pollen, as pollinator reward, 113–114, 139–140
Pollination, 111, 113–114, 118, 120–122, 137–158
Pollinia, 145, 149
Polygonum, 193
Polysporangiophytes, 66, 74
Polytrichum, 49, 69
Pontederia, Pontederiaceae, 219, 222, 223, 225
Populus tremuloides, 159, 174
Potentilla, 140
Prickle, 185, 186
Primary root, 92, 101, 168
Primary thickening meristem. See under meristem
Prochlorophytes, 28, 30
Progymnosperms, 89, 91–93, 99, 102–104
Proteaceae, 146, 153
Protists, 22–23, 27, 28, 31, 38, 191–192
Proton pump, motive force, 9–10
Protozoa, 22
Prunus, 140, 162
Pseudobulb, 213
Pseudocopulation, 149
Pseudostem, 208
Psilotales, 84–85
Psilotum, 83, 85, 86
Pueraria lobata, 180
Purple non-sulfur bacteria, 13

Pyrostegia ignea, 198

Quercus, 163, 164, 184

Radial symmetry, 87, 92, 100–101
Rafflesia, 147, *148*
Ravenala madagascariensis, 147
Ray flowers, 153–154
Rays, in wood, 169, *170*
Receptacle, 162, *163*
Red algae, see algae
Reduction (redox) potential, 7, 12
Resurrection fern, 56, 57
Reticulate venation, 181–*182*
  See also net venation
Rhizobium, 14, 15
Rhizoids, 49, 68, 74
Rhizome, 67, 74, 75, 77, 80, 82, 84, 85, 87, 89–90, 92, 101, 158–159, 172, 177, *179*, 192, 193, 196, 199, *200*, 206, 210, *211*, 213, *214*, 215, *217*, 219, *221*, 224
Rhizophora, 173
Rhynia, 75, 84
Root crown, 175, 177
Root system, axial, 91–93, 100–101, 168
Root system, adventitious, 91, 168, 172–*174*, 177, *200*, *201*, *211*, 213, 215, *216*, 224
Roots, 68, 73, 77, 78, *101*
  tuberous, 175, 177, *180*
  See also adventitious roots; taproot
Root-stem axis, 91–93, 101, 172–173
Roridula, 189
Rosette, 87, 88, *171*, 172, 175–*176*, 215, 218
Rosette tree, 184, *185*
Rubisco, 7
Runner, 175

Sagittaria, 218, 223, 224
Salvinia, 83, 98
Sansevieria, 216
Sarracenia, 189, *190*
Saururus, 192
Savannas, African, *198*, 199–202, 216
Sclerophyll, 188
Scrophulariaceae, 140
Sea slugs, photosynthetic, *16*, 17–18
Sea grasses, 158, 201, 215, 218, 219, 221, 224
Seaweeds, 22
Secondary growth, 171, 173, 174, 215, 218
Secondary plant compounds, 195–196
Secondary xylem, 73, 89
Seeds, 94–97
Seed ferns, 101, *102*, 103–104, 110, 113, 127
  Mesozoic, 128
Selaginella, 96, 97, 98, *101*
Self-pollination, avoidance, 150, 156–157
Senecio, 188

Sequoia sempervirens, 73, 106
Sequoiadendron giganteum, 72, 73
Serenoa, 215
Sexual reproduction, 32, *36*, *38*, 51–52, 60–63, 96, 97, 98
Sigillaria, 89, *90*
Silurian period, 74
Smilax, 85
Sperm, 32, 38–39, 60–61, 74, 96, 97, 98, 99, 100
Spermatia of red algae, 31
Sphagnum, 47, *49*, 53, *54*, 55, 65
Sphenophyllum, 82, *85*
Sphenophyte, 81–82
Spirochaete, 23
Spirogyra, 39, *40*, *41*
Splachnum, 65
Sporangium, 63–64, *65*, 66–68, 76, 99–100
Spore dispersal, 31, 35, 37, 52, 62–65, 79
Spores, xv, 31, 38, 52, 62–63, *64*, 65, 86, 95, 96, 97–99, *100*, 113
  See also zoospores
Sporophyte, *38*, 52
Sporophyte, in green algae, 37
Sporophytes of vascular plants, 60–63, *64*, 65, 66–68, 70
Sporopollenin, 62
Stamens, *120*, *121*, 122–123, *124*, 127, 129, 134, 144, 156, *158*
Stapelia, 148, *187*, 188
Stem group, angiosperm, *119*, 120, 128, 135, *133*
Stigma, *120*, *121*, 123, 125, 131, 132, 134, 135, 139, *144*, 150, 156–157, *158*, 236
Stolon, 175, *176*, 177
Stomata, *58*, 188, 217
Strobilus, 79, *80*
Stromatolites, *xviii*, 1–2, 5, 235
Succulent, stems and leaves, 116, *187*, 188, 201, 213, 216, *219*
Sunbirds, as pollinators, 145
Sundew, 172, 190, *191*
Suture, in angiosperm carpel, *124*, 131–132, *133*, 134–135
Symmetry in plants, radial and bilateral, 87, 92, 100–101, 133, 144, *153*, 171–172, 199

Taproot, 108, *109*, 168, *171*, 172, 199
Taxodium, 106
Taxus, 95, 135, 160
Tendrils, *181*, 184–185
Tepals, *120*, 123, 127, *149*
Thalassia, 158
Thallus, thalloid growth form, 44, *48*, *49*, 51, *52*, 61
Thermoacidophiles, 20
Thorn, 185, *186*
Thylakoids, 9, *10*, *11*
Tmesipteris, 86

Tofieldia, 226, 228
Tracheids, 58, *59*, 60, 75, 89, 91, 110, 132, 169, *170*, 171
Tracheophyta, 48
Transpiration, 55, 58, 60, 91
Trap flowers, 122, 155–156
Trapa, 193
Trees, evolution of, 66, 73, 85, 88–89, *90*, *91*, 92–93, 100–101
Trees, in monocots, 201–202, 208, 213, *214*, 215, *216*, *218*
Trewia nudiflora, 162
Tubers, 175, 177
Tuberous roots, 175, 177, *180*
Turgor pressure, turgid, 53, 55, 59
Typha, Typhaceae, 205, 219

Ulothrix, 38, 40, *41*
Ultraviolet and pollination, 143
Ultraviolet radiation, 8, 50
Ulva, 36, *37*, 38, 40, 60, 64, 67–68
Unifacial leaf, 228
Unisexual flowers, 122, 132, 150, 156
Utricularia, 191, *192*, 193

Vallisneria, 158, *159*, 219, *221*
Vascular cambium, 89–90
Vascular plants, xv, 48, 51, 56–60, 66–69
Vascular tissues, 48, 60, 70, 75, 168, *169*, 181
 secondary, 89
 See also xylem; phloem; tracheid; vessel

Venus flytrap, *171*, 172
Vessels, 110, 132, 169, *170*, 171
Vines, 175, 176, 180, *181*, 185, 205
Volvox, 38, *39*, 40

Warrawoona cherts, 2
Wasps, as pollinators, *149*, 154
Water, in pollination, 157–158, *159*
Water lilies, 120–122, 139, 155, 168, 177, 192, 193, *194*, 218, 224
Welwitschia, 94, 108, *109*
Wind, in seed dispersal, 159, *160*, 163, *164*
Wind, in pollination, 156, *157*, 158
Winteraceae, 132, 170–171
Wood, 73, 88–89, 132, 168, 169, *170*, 171

Xanthorrhea, 216
Xerophytes, in eudicots, *187*, 216, *219*
Xylem, 59, 66, 75, *169*
Xylem, secondary, 73, 89, *90*, 170, 215

Yucca, 155, 216, *219*

Zamia, Zamiaceae, 111, *112*
Zingiberaceae, 196, 206, 212
Zoospores, 33–36, *37*, 60, 62–64, 68
Zostera, 158
Zosterophyllum, 75, 76, 77
Zygomorphic, 144
Zygote, 32–33, 35, *36*, *37*, *38*, *40*, 52, 62–63, 67–68